Microbiology

by
I. Edward Alcamo, Ph.D.

INCORPORATED

LINCOLN, NEBRASKA 68501

Illustrations by Thomas J. Gagliano/Gagliano Graphics

Cover photograph by Phil Degginger/Tony Stone Images

FIRST EDITION

ISBN 0-8220-5333-0

Microorganisms surround us on all sides and make their presence felt throughout our lives. The useful species outnumber the harmful ones by thousands to one and are so valuable that life as we know it would be impossible without them. Microorganisms break down the remains of all that die and recycle such elements as carbon, nitrogen, and sulfur; they produce numerous foods, including cheeses, dairy products, and fermented beverages; they are essential decomposers in sewage treatment; and they manufacture numerous industrial products such as drugs, insecticides, hormones, and enzymes.

Unfortunately, microorganisms also cause human disease. Multiplying in their countless numbers in the tissues, they produce toxins that affect body systems and destroy the body by their sheer force of numbers. Microorganisms have been involved in the great plagues of history, and their effects have captured the imaginations of scientists, writers, and chroniclers of society. Indeed, untold research dollars have been spent searching for ways to stop epidemics in the environment and enhance the body's natural defenses.

The microbial world is fascinating to explore. It includes the microscopic bacteria, fungi, and protozoa; the submicroscopic rickettsiae and chlamydiae; and the ultramicroscopic viruses. In *Cliffs Quick Review Microbiology*, we summarize the characteristics of these microorganisms and examine their myriad activities in the spectrum of life. A major theme is microbial disease and the means available to protect against it both inside and outside the body.

Cliffs Quick Review Microbiology contains the foundation material for microbiology courses taken by health and allied health science students. Students preparing for careers in nursing, dental hygiene, medical technology, food and nutrition, and pharmacy will find the book useful. Premedical and medical students can use it as an adjunct in their professional courses. And advanced high school students will benefit from its direct approach to the microorganisms.

We have presented the content material of *Cliffs Quick Review Microbiology* in clear, concise detail that can be read easily and understood immediately. The paragraphs are brief, and the essential vocabulary and key topic areas have been included. Indeed, the 29 chapters conform to the chapters of the major microbiology texts available today.

Cliffs Quick Review Microbiology has benefited from the talents of many individuals, and I would like to pause and thank them. I am pleased to acknowledge the expert guidance of editor Michele Spence during the development of this book. I also express my gratitude to Barbara Dunleavy, who typed the manuscript carefully and efficiently. And I am happy to thank the many professionals who lent their expertise during the production phase of this book. Readers who wish to reach me with their suggestions and comments may write me in care of Cliffs Notes, Inc.

In closing, I want to extend to you, the student, my best wishes for a successful experience in microbiology.

E. Alcamo
State University of New York
Farmingdale, NY 11735

DISEASES OF THE RESPIRATORY SYSTEM ... 195

Microorganisms are a collection of organisms that share the characteristic of being visible only with a microscope. They constitute the subject matter of **microbiology.**

Members of the microbial world are very diverse and include the bacteria, cyanobacteria, rickettsiae, chlamydiae, fungi, unicellular (single-celled) algae, protozoa, and viruses. The majority of microorganisms contribute to the quality of human life by doing such things as maintaining the balance of chemical elements in the natural environment, by breaking down the remains of all that dies, and by recycling carbon, nitrogen, sulfur, phosphorus, and other elements.

Some species of microorganisms cause infectious disease. They overwhelm body systems by sheer force of numbers, or they produce powerful toxins that interfere with body physiology. Viruses inflict damage by replicating within tissue cells, thereby causing tissue degeneration.

The Spectrum of Microbiology

Like all other living things, microorganisms are placed into a system of **classification.** Classification highlights characteristics that are common among certain groups while providing order to the variety of living things. The science of classification is known as **taxonomy, and taxon** is an alternative expression for a classification category. Taxonomy displays the unity and diversity among living things, including microorganisms. Among the first taxonomists was **Carolus Linnaeus.** In the 1750s and 1760s, Linnaeus classified all known plants and animals of that period and set down the rules for nomenclature.

Classification schemes. The fundamental rank of the classification as set down by Linnaeus is the **species.** For organisms such as animals and plants, a species is defined as a population of individuals that breed among themselves. For microorganisms, a species is defined as a group of organisms that are 70 percent similar from a biochemical standpoint.

In the classification scheme, various species are grouped together to form a **genus.** Among the bacteria, for example, the species *Shigella boydii* and *Shigella flexneri* are in the genus *Shigella* because the organisms are at least 70 percent similar. Various genera are then grouped as a **family** because of similarities, and various families are placed together in an **order.** Continuing the classification scheme, a number of orders are grouped as a **class,** and several classes are categorized in a single **phylum** or **division.** The various phyla or divisions are placed in the broadest classification entry, the **kingdom.**

Numerous criteria are used in establishing a species and in placing species together in broader classification categories. Morphology (form) and structure are considered, as well as cellular features, biochemical properties, and genetic characteristics. In addition, the antibodies that an organism elicits in the human body are a defining property. The nutritional format is considered, as are staining characteristics.

Prokaryotes and eukaryotes. Because of their characteristics, microorganisms join all other living organisms in two major groups of organisms: prokaryotes and eukaryotes. Bacteria are **prokaryotes** (simple organisms having no nucleus or organelles) because of their cellular properties, while other microorganisms such as fungi, protozoa, and unicellular algae are **eukaryotes** (more complex organisms whose cells have a nucleus and organelles). Viruses are neither prokaryotes nor eukaryotes because of their simplicity and unique characteristics.

The five kingdoms. The generally accepted classification of living things was devised by **Robert Whittaker** of Cornell University in 1969. Whittaker suggested a five-kingdom classification.

The first of the five kingdoms is **Monera** (in some books, Prokaryotae). Prokaryotes, such as bacteria and cyanobacteria (formerly, blue-green algae), are in this kingdom; the second kingdom, **Protista,** includes protozoa, unicellular algae, and slime molds, all of which are eukaryotes and single-celled; in the third kingdom, **Fungi,** are the molds, mushrooms, and yeasts. These organisms are eukaryotes that absorb simple nutrients from the soil (Figure 1). The remaining two kingdoms are **Plantae** (plants) and **Animalia** (animals).

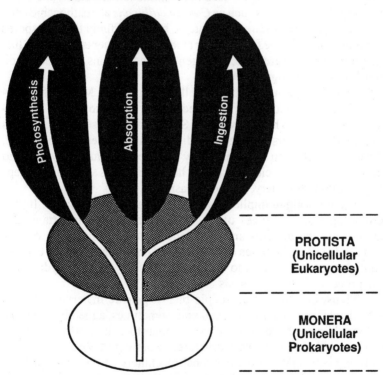

The five kingdoms of living things showing the position of microorganisms relative to other organisms.

■ Figure 1 ■

Brief descriptions of microorganisms. **Bacteria** are relatively simple, prokaryotic organisms whose cells lack a nucleus or nuclear membrane. The bacteria may appear as rods (bacilli), spheres (cocci), or spirals (spirilla or spirochetes). Bacteria reproduce by binary fission, have unique constituents in their cell walls, and exist in most environments on earth. For instance, they live at temperatures ranging from 0° to 100°C and in conditions that are oxygen rich or oxygen free. A microscope is necessary to see and study them.

Fungi are eukaryotic microorganisms that include multicellular molds and unicellular (single-celled) yeasts. The **yeasts** are slightly larger than bacteria and are used in alcoholic fermentations and bread making. Certain yeasts such as *Candida albicans* are pathogenic (disease causing). **Molds** are filamentous, branched fungi that use spores for reproduction. The fungi prefer acidic environments, and most live at room temperature under oxygen-rich conditions. The common mushroom is a fungus.

Protozoa are eukaryotic, unicellular organisms. Motion is a characteristic associated with many species, and the protozoa can be classified according to how they move: Some protozoa use flagella, others use cilia, and others use pseudopodia. Certain species are nonmotile. Protozoa exist in an infinite variety of shapes because they have no cell walls. Many species cause such human diseases as malaria, sleeping sickness, dysentery, and toxoplasmosis.

The term **algae** implies a variety of plantlike organisms. In microbiology, several types of single-celled algae are important. Examples are the diatoms and dinoflagellates that inhabit the oceans and are found at the bases of marine food chains. Most algae capture sunlight and transform it to the chemical energy of carbohydrates in the process of photosynthesis.

Viruses are ultramicroscopic bits of genetic material (DNA or RNA) enclosed in a protein shell and, sometimes, a membranous envelope. Viruses have no metabolism; therefore, it is difficult to use drugs to interfere with their structures or activities. Viruses multiply in living cells and use the chemical machinery of the cells for their own purpose. Often, they destroy the cells in the process of replicating.

Nomenclature of microorganisms. The system for naming all living things, established by Linnaeus, is also applied to microorganisms. In this system, all organisms are placed into a classification system, and each organism is given a binomial name. The **binomial name** consists of two names. The first name is the **genus** to which the organism belongs. The second name is a modifying adjective called the **species modifier.**

In writing the binomial name, the first letter of the genus name is capitalized, and the remainder of the genus name and the complete species modifier are written in lowercase letters. The entire binomial name is either italicized or underlined. It can be abbreviated by using the first letter of the genus name and the full species modifier. An example of a microbial name is *Escherichia coli*, the bacterial rod found in the human intestine. The name is abbreviated *E. coli.*

A Brief History of Microbiology

Microbiology has had a long, rich history, initially centered in the causes of infectious diseases but now including practical applications of the science. Many individuals have made significant contributions to the development of microbiology.

Early history of microbiology. Historians are unsure who made the first observations of microorganisms, but the microscope was available during the mid-1600s, and an English scientist named **Robert Hooke** made key observations. He is reputed to have observed strands of fungi among the specimens of cells he viewed. In the 1670s and the decades thereafter, a Dutch merchant named **Anton van Leeuwenhoek** made careful observations of microscopic organisms, which he called **animalcules.** Until his death in 1723, van Leeuwenhoek revealed the microscopic world to scientists of the day and is regarded as one of the first to provide accurate descriptions of protozoa, fungi, and bacteria.

After van Leeuwenhoek died, the study of microbiology did not develop rapidly because microscopes were rare and the interest in microorganisms was not high. In those years, scientists debated the theory of **spontaneous generation,** which stated that microorganisms arise from lifeless matter such as beef broth. This theory was disputed by **Francesco Redi,** who showed that fly maggots do not arise from decaying meat (as others believed) if the meat is covered to prevent the entry of flies. An English cleric named **John Needham** advanced spontaneous generation, but **Lazzaro Spallanzani** disputed the theory by showing that boiled broth would not give rise to microscopic forms of life.

Louis Pasteur and the germ theory. Louis Pasteur worked in the middle and late 1800s. He performed numerous experiments to discover why wine and dairy products became sour, and he found that bacteria were to blame. Pasteur called attention to the importance of microorganisms in everyday life and stirred scientists to think that if bacteria could make the wine "sick," then perhaps they could cause human illness.

Pasteur had to disprove spontaneous generation to sustain his theory, and he therefore devised a series of **swan-necked flasks** filled with broth. He left the flasks of broth open to the air, but the flasks had a curve in the neck so that microorganisms would fall into the neck, not the broth. The flasks did not become contaminated (as he predicted they would not), and Pasteur's experiments put to rest the notion of spontaneous generation. His work also encouraged the belief that microorganisms were in the air and could cause disease. Pasteur postulated the **germ theory of disease,** which states that microorganisms are the causes of infectious disease.

Pasteur's attempts to prove the germ theory were unsuccessful. However, the German scientist **Robert Koch** provided the proof by cultivating anthrax bacteria apart from any other type of organism. He then injected· pure cultures of the bacilli into mice and showed that the bacilli invariably caused anthrax. The procedures used by Koch came to be known as **Koch's postulates** (Figure 2). They provided a set of principles whereby other microorganisms could be related to other diseases.

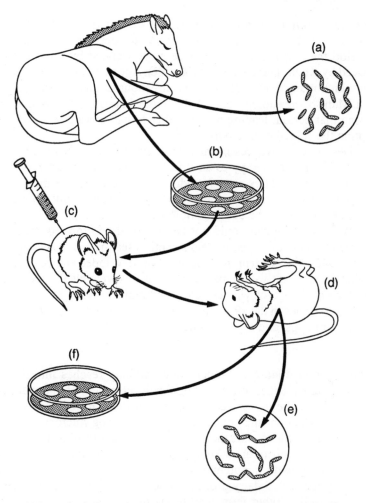

The steps of Koch's postulates used to relate a specific microorganism to a specific disease. (a) Microorganisms are observed in a sick animal and (b) cultivated in the lab. (c) The organisms are injected into a healthy animal, and (d) the animal develops the disease. (e) The organisms are observed in the sick animal and (f) reisolated in the lab.

■ Figure 2 ■

The development of microbiology. In the late 1800s and for the first decade of the 1900s, scientists seized the opportunity to further develop the germ theory of disease as enunciated by Pasteur and proved by Koch. There emerged a **Golden Age of Microbiology** during which many agents of different infectious diseases were identified. Many of the etiologic agents of microbial disease were discovered during that period, leading to the ability to halt epidemics by interrupting the spread of microorganisms.

Despite the advances in microbiology, it was rarely possible to render life-saving therapy to an infected patient. Then, after World War II, the **antibiotics** were introduced to medicine. The incidence of pneumonia, tuberculosis, meningitis, syphilis, and many other diseases declined with the use of antibiotics.

Work with viruses could not be effectively performed until instruments were developed to help scientists see these disease agents. In the 1940s, the **electron microscope** was developed and perfected. In that decade, cultivation methods for viruses were also introduced, and the knowledge of viruses developed rapidly. With the development of vaccines in the 1950s and 1960s, such viral diseases as polio, measles, mumps, and rubella came under control.

Modern microbiology. Modern microbiology reaches into many fields of human endeavor, including the development of pharmaceutical products, the use of quality-control methods in food and dairy product production, the control of disease-causing microorganisms in consumable waters, and the industrial applications of microorganisms. Microorganisms are used to produce vitamins, amino acids, enzymes, and growth supplements. They manufacture many foods, including fermented dairy products (sour cream, yogurt, and buttermilk), as well as other fermented foods such as pickles, sauerkraut, breads, and alcoholic beverages.

One of the major areas of applied microbiology is **biotechnology.** In this discipline, microorganisms are used as living factories to produce pharmaceuticals that otherwise could not be manufactured. These substances include the human hormone insulin, the antiviral substance interferon, numerous blood-clotting factors and clot-dissolving enzymes, and a number of vaccines. Bacteria can be reengineered to increase plant resistance to insects and frost, and biotechnology will represent a major application of microorganisms in the next century.

In the 1700s, scientists discovered the chemical and physical basis of living things, and soon they realized that the chemical organization of all living things is remarkably similar. Microorganisms, as forms of living things, conform to this principle and have a chemical basis that underlies their metabolism.

Chemical Principles

Elements and atoms. All living things on earth, including microorganisms, are composed of fundamental building blocks of matter called **elements.** Over 100 elements are known to exist, including certain ones synthesized by scientists. An element is a substance that cannot be decomposed by chemical means. Such things as oxygen, iron, calcium, sodium, hydrogen, carbon, and nitrogen are elements.

Each element is composed of one particular kind of atom. An **atom** is the smallest part of an element that can enter into combinations with atoms of other elements.

Atoms consist of positively charged particles called **protons** surrounded by negatively charged particles called **electrons.** A third particle called the **neutron** has no electrical charge; it has the same weight as a proton. Protons and neutrons adhere tightly to form the dense, positively charged **nucleus** of the atom. Electrons spin around the nucleus in orbits, or shells.

The arrangement of electrons in an atom plays an essential role in the chemistry of the atom. Atoms are most stable when their outer shell of electrons has a full quota, which may be two electrons or eight electrons. Atoms tend to gain or lose electrons until their outer shells have this stable arrangement. The gaining or losing of electrons contributes to the chemical reactions in which an atom participates.

Molecules. Most of the microbial compounds of interest to biologists are composed of units called molecules. A **molecule** is a precise arrangement of atoms from different elements; a **compound** is a mass of molecules . The arrangements of the atoms in a molecule account for the properties of a compound. The molecular weight is equal to the atomic weights of the atoms in the molecule. For example, the molecular weight of water is 18.

The atoms in molecules may be joined to one another by various linkages called bonds. One example of a bond is an **ionic bond,** which is formed when the electrons of one atom transfer to a second atom, creating electrically charged atoms called **ions.** The electrical charges attract the ions to one another; the attraction creates the ionic bond. Sodium chloride consists of sodium ions and chloride ions joined by ionic bonds (Figure 3a).

A second type of linkage is called a **covalent bond** (Figure 3b), which forms when two atoms share one or more electrons with one another. For example, carbon shares its electrons with four hydrogen atoms, and the resulting molecule is methane (CH_4). If one pair of electrons is shared, the bond is a single bond; if two pairs are shared, then it is a double bond. Covalent bonds are present in organic molecules such as proteins, lipids, and carbohydrates.

Acids and bases. Certain chemical compounds release hydrogen ions when the compounds are placed in water. These compounds are called **acids.** For example, when hydrogen chloride is placed in water, it releases its hydrogen ions, and the solution becomes hydrochloric acid.

Certain chemical compounds attract hydrogen ions when they are placed in water. These substances are called **bases.** An example of a base is sodium hydroxide (NaOH). When this substance is placed in water, it attracts hydrogen ions, and a basic (or alkaline) solution results.

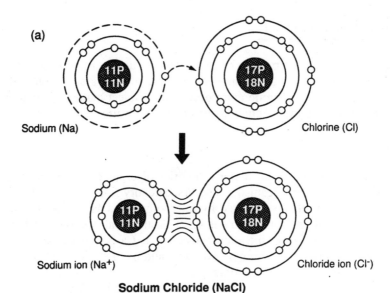

(a)

Sodium (Na)

Chlorine (Cl)

Sodium ion (Na+)

Chloride ion (Cl-)

Sodium Chloride (NaCl)

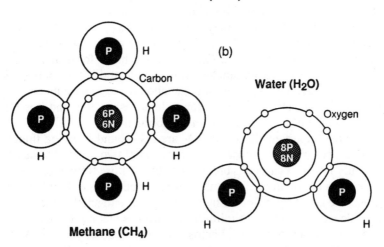

P   H

Carbon

(b)

Water (H2O)

P   6P
          6N   P   Oxygen

H   H   8P
                                8N

P   H   P   P

H   H

Methane (CH4)

*Bond formation in molecules. (a) Formation of an ionic bond in
a sodium chloride molecule. (b) Covalent bonding in methane
and water molecules.*

■ Figure 3 ■

Organic Compounds

The chemical compounds of living things such as microorganisms are known as organic compounds because of their association with organisms. The **organic compounds,** the subject matter of organic chemistry, are the compounds associated with life processes in microorganisms.

Carbohydrates. Four major categories of organic compounds are found in all microorganisms. The first category is the carbohydrates.

Carbohydrates are used by microorganisms as sources of energy. In addition, carbohydrates serve as structural materials such as in the construction of the microbial cell wall. Carbohydrates are molecules composed of carbon, hydrogen, and oxygen; the ratio of hydrogen atoms to oxygen atoms is 2:1.

The simple carbohydrates are commonly referred to as sugars. Sugars are **monosaccharides** if they are composed of single molecules and disaccharides if they are composed of two molecules. The most important monosaccharide is glucose, a carbohydrate with the molecular formula $C_6H_{12}O_6$. Glucose is the basic form of fuel for many species of microorganisms. It is soluble and is transported by body fluids to all cells, where it is metabolized to release its energy. Glucose is the starting material for cellular respiration, and it is the main product of photosynthesis in microorganisms.

Three important **disaccharides** are also found in living things. One disaccharide is maltose, a combination of two glucose units covalently linked. Yeast cells break down the maltose from grain starch in the process of alcoholic fermentation. Another disaccharide is sucrose, the table sugar formed by linking glucose to another monosaccharide called fructose. A third disaccharide is lactose, composed of glucose and galactose units. Lactose, the major carbohydrate in milk, is digested to acid by microorganisms when they sour milk and form sour-milk products such as yogurt and sour cream.

Complex carbohydrates are known as **polysaccharides.** Polysaccharides are formed by linking eight or more monosaccharide molecules. Among the most important polysaccharides are **starches,** which are composed of hundreds or thousands of glucose units linked to one another. Starches serve as a storage form for carbohydrates. Microorganisms break down starch to use the glucose it contains for their energy needs.

Another important polysaccharide is **glycogen,** which is related to starch. Many bacteria have glycogen in thier cytoplasm. Still another is **cellulose.** Cellulose is also composed of glucose units, but the units cannot be released from one another except by a few species of microorganisms, especially those in the stomach of the cow and other ruminants. The cell walls of algae contain cellulose, and certain fungi have this polysaccharide. Another polysaccharide called **chitin** is a primary constituent in the fungal cell wall.

Lipids. **Lipids** are organic molecules composed of carbon, hydrogen, and oxygen atoms. In contrast to carbohydrates, the ratio of hydrogen atoms to oxygen atoms is much higher. Lipids include steroids, waxes, and the most familiar lipids, fats.

Fat molecules are composed of a glycerol molecule and one, two, or three molecules of fatty acids. A fatty acid is a long chain of carbon atoms with associated hydroxyl (–OH) groups. At one end of the fatty acid is an organic acid (–COOH) group . The fatty acids in a fat may be all alike or all different. They are bound to the glycerol molecule during **dehydration synthesis,** a process that involves the removal of water (Figure 4a). The number of carbon atoms in a fatty acid may be as few as four or as many as 24.

Certain fatty acids have one or more double bonds in their molecules. Fats that include these molecules are called **unsaturated fats.** Other fatty acids have no double bonds. Fats that include these fatty acids are called **saturated fats.**

Some microbial species use fats as energy sources. They produce the enzyme lipase, which breaks down fats to fatty acids and glycerol. An important type of phosphorus-containing lipid, the **phospholipid,** is a major constituent of the cell membranes of all microorganisms.

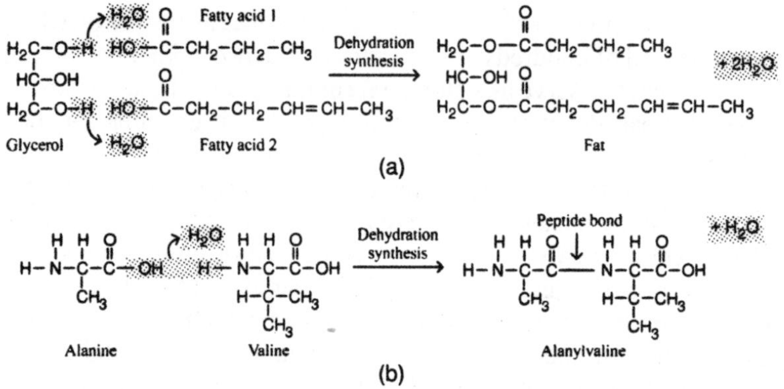

Syntheses in organic molecules. (a) Bonding of two fatty acids
to a glycerol molecule in the formation of a fat. (b) Bonding of
two amino acids via a peptide bond in the formation of a protein.

■ Figure 4 ■

Proteins. **Proteins** are among the most complex of all organic compounds. They are composed of units called **amino acids,** which contain carbon, hydrogen, oxygen, and nitrogen atoms. Certain amino acids also have sulfur atoms, phosphorus, or other trace elements such as iron or copper.

Many proteins are immense and complex as compared to carbohydrates or fats. However, all are composed of folded, long chains of the relatively simple amino acids. There are 20 kinds of amino acids, each with an amino ($-NH_2$) group and an organic acid ($-COOH$) group. The amino acids differ with respect to the nature of the chemical group that is attached to the base structure. Examples of amino acids are alanine, valine, glutamic acid, tryptophan, tyrosine, and histidine.

Amino acids are linked to form a protein by the removal of water molecules (Figure 4b). The links forged between the amino acids are called **peptide bonds,** and small proteins are often called **peptides.**

All living things, including microorganisms, depend upon proteins for their existence. Proteins are the major molecules from which microorganisms are constructed. Certain proteins are dissolved or suspended in the watery substance of the cells, while others are incorporated into various structures of the cells, such as the cell membrane. Bacterial toxins (metabolic poisons) and microbial flagella and pili are usually composed of proteins.

An essential use for proteins is in the construction of enzymes. **Enzymes** catalyze the chemical reactions that take place within microorganisms. The enzymes are not used up in the reaction, but remain available to catalyze succeeding reactions. Without enzymes, the metabolic activity of the microorganism could not take place.

Every species manufactures proteins unique to that species. The information for synthesizing these unique proteins is found in the nucleus of the cell. The so-called **genetic code** specifies the sequence of amino acids in the protein and thereby regulates the chemical activity taking place within the cell. Proteins also can serve as a reserve source of energy for the microorganism. When the amino group is removed from an amino acid, the resulting compound is energy rich.

Nucleic acids. Like proteins, **nucleic acids** are very large molecules. The nucleic acids are composed of smaller units called **nucleotides.** Each nucleotide contains a five-carbon carbohydrate molecule, a phosphate group, and a nitrogen-containing molecule that has basic properties and is called a nitrogenous base.

Microorganisms contain two important kinds of nucleic acids. One type is called **deoxyribonucleic acid,** or **DNA.** The other is known as **ribonucleic acid,** or **RNA.** DNA is found primarily in the nucleus of eukaryotic microorganisms (which have nuclei) and suspended in the cytoplasm of prokaryotic microorganisms (which lack nuclei). DNA is also located in plasmids, the tiny loops of DNA found in bacterial cytoplasm. RNA is found in both the nucleus (if present) and the cytoplasm of the microorganism.

DNA and RNA differ from one another in their components. DNA contains the carbohydrate deoxyribose, while RNA has ribose. In addition, DNA contains the bases adenine, cytosine, guanine, and thymine, while RNA has adenine, cytosine, guanine, and uracil.

Since microorganisms are invisible to the unaided eye, the essential tool in microbiology is the microscope. One of the first to use a microscope to observe microorganisms was **Robert Hooke,** the English biologist who observed algae and fungi in the 1660s. In the 1670s, **Anton van Leeuwenhoek,** a Dutch merchant, constructed a number of simple microscopes and observed details of numerous forms of protozoa, fungi, and bacteria. During the 1700s, microscopes were used to further elaborate on the microbial world, and by the late 1800s, the sophisticated light microscopes had been developed. The electron microscope was developed in the 1940s, thus making the viruses and the smallest bacteria (for example, rickettsiae and chlamydiae) visible.

Microscopes permit extremely small objects to be seen, objects measured in the metric system in micrometers and nanometers. A **micrometer** (μm) is equivalent to a millionth of a meter, while a **nanometer** (nm) is a billionth of a meter. Bacteria, fungi, protozoa, and unicellular algae are normally measured in micrometers, while viruses are commonly measured in nanometers. A typical bacterium such as *Escherichia coli* measures about two micrometers in length and about one micrometer in width.

Types of Microscopes

Various types of microscopes are available for use in the microbiology laboratory. The microscopes have varied applications and modifications that contribute to their usefulness.

The light microscope. The common light microscope used in the laboratory is called a **compound microscope** because it contains two types of lenses that function to magnify an object. The lens closest to the eye is called the **ocular,** while the lens closest to the object is called the **objective.** Most microscopes have on their base an apparatus called a **condenser,** which condenses light rays to a strong beam. A **diaphragm** located on the condenser controls the amount of light coming through it. Both coarse and fine adjustments are found on the light microscope (Figure 5a).

To magnify an object, light is projected through an opening in the stage, where it hits the object and then enters the objective. An image is created, and this image becomes an object for the ocular lens, which remagnifies the image. Thus, the **total magnification** possible with the microscope is the magnification achieved by the objective multiplied by the magnification achieved by the ocular lens.

A compound light microscope often contains four **objective lenses:** the scanning lens (4×), the low-power lens (10×), the high-power lens (40×), and the oil-immersion lens (100×). With an ocular lens that magnifies 10 times, the total magnifications possible will be 40× with the scanning lens, 100× with the low-power lens, 400× with the high-power lens, and 1000× with the oil-immersion lens. Most microscopes are **parfocal.** This term means that the microscope remains in focus when one switches from one objective to the next objective.

The ability to see clearly two items as separate objects under the microscope is called the **resolution** of the microscope. The resolution is determined in part by the wavelength of the light used for observing. Visible light has a wavelength of about 550 nm, while ultraviolet light has a wavelength of about 400 nm or less. The resolution of a microscope increases as the wavelength decreases, so ultraviolet light allows one to detect objects not seen with visible light. The **resolving power** of a lens refers to the size of the smallest object that can be seen with that lens. The resolving power is based on the wavelength of the light used and the numerical aperture of the lens. The **numerical aperture** (NA) refers to the widest cone of light that can enter the lens; the NA is engraved on the side of the objective lens.

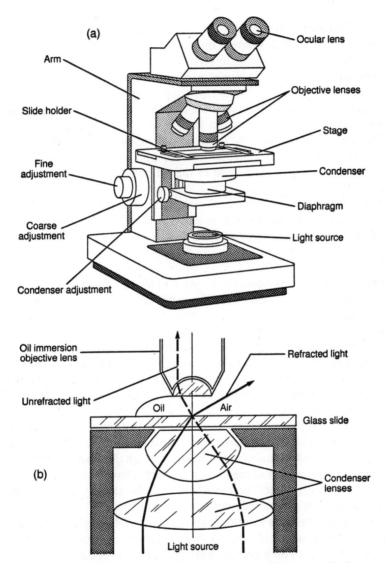

Light microscopy. (a) The important parts of a common light microscope. (b) How immersion oil gathers more light for use in the microscope.

■ Figure 5 ■

If the user is to see objects clearly, sufficient light must enter the objective lens. With modern microscopes, entry to the objective is not a problem for scanning, low-power, and high-power lenses. However, the oil-immersion lens is exceedingly narrow, and most light misses it. Therefore, the object is seen poorly and without resolution. To increase the resolution with the oil-immersion lens, a drop of **immersion oil** is placed between the lens and the glass slide (Figure 5b). Immersion oil has the same light-bending ability (index of refraction) as the glass slide, so it keeps light in a straight line as it passes through the glass slide to the oil and on to the glass of the objective, the oil-immersion lens. With the increased amount of light entering the objective, the resolution of the object increases, and one can observe objects as small as bacteria. Resolution is important in other types of microscopy as well.

Other light microscopes. In addition to the familiar compound microscope, microbiologists use other types of microscopes for specific purposes. These microscopes permit viewing of objects not otherwise seen with the light microscope.

An alternative microscope is the **dark-field microscope,** which is used to observe live spirochetes, such as those that cause syphilis. This microscope contains a special condenser that scatters light and causes it to reflect off the specimen at an angle. A light object is seen on a dark background.

A second alternative microscope is the **phase-contrast microscope.** This microscope also contains special condensers that throw light "out of phase" and cause it to pass through the object at different speeds. Live, unstained organisms are seen clearly with this microscope, and internal cell parts such as mitochondria, lysosomes, and the Golgi body can be seen with this instrument.

The **fluorescent microscope** uses ultraviolet light as its light source. When ultraviolet light hits an object, it excites the electrons of the object, and they give off light in various shades of color. Since ultraviolet light is used, the resolution of the object increases. A laboratory technique called the fluorescent-antibody technique employs fluorescent dyes and antibodies to help identify unknown bacteria.

Electron microscopy. The energy source used in the **electron microscope** is a beam of electrons. Since the beam has an exceptionally short wavelength, it strikes most objects in its path and increases the resolution of the microscope significantly. Viruses and some large molecules can be seen with this instrument. The electrons travel in a vacuum to avoid contact with deflecting air molecules, and magnets focus the beam on the object to be viewed. An image is created on a monitor and viewed by the technologist.

The more traditional form of electron microscope is the **transmission electron microscope (TEM)**. To use this instrument, one places ultrathin slices of microorganisms or viruses on a wire grid and then stains them with gold or palladium before viewing. The densely coated parts of the specimen deflect the electron beam, and both dark and light areas show up on the image.

The **scanning electron microscope (SEM)** is the more contemporary form of electron microscope. Although this microscope gives lower magnifications than the TEM, the SEM permits three-dimensional views of microorganisms and other objects. Whole objects are used, and gold or palladium staining is employed.

Staining Techniques

Because microbial cytoplasm is usually transparent, it is necessary to stain microorganisms before they can be viewed with the light microscope. In some cases, staining is unnecessary, for example when microorganisms are very large or when motility is to be studied, and a drop of the microorganisms can be placed directly on the slide and observed. A preparation such as this is called a **wet mount**. A wet mount can also be prepared by placing a drop of culture on a coverslip (a glass cover for a slide) and then inverting it over a hollowed-out slide. This procedure is called the **hanging drop**.

In preparation for staining, a small sample of microorganisms is placed on a slide and permitted to air dry. The smear is heat fixed by quickly passing it over a flame. **Heat fixing** kills the organisms, makes them adhere to the slide, and permits them to accept the stain.

Simple stain techniques. Staining can be performed with basic dyes such as crystal violet or methylene blue, positively charged dyes that are attracted to the negatively charged materials of the microbial cytoplasm. Such a procedure is the **simple stain procedure.** An alternative is to use a dye such as nigrosin or Congo red, acidic, negatively charged dyes. They are repelled by the negatively charged cytoplasm and gather around the cells, leaving the cells clear and unstained. This technique is called the **negative stain technique.**

Differential stain techniques. The **differential stain technique** distinguishes two kinds of organisms. An example is the **Gram stain technique.** This differential technique separates bacteria into two groups, Gram-positive bacteria and Gram-negative bacteria. Crystal violet is first applied, followed by the mordant iodine, which fixes the stain (Figure 6). Then the slide is washed with alcohol, and the Gram-positive bacteria retain the crystal-violet iodine stain; however, the Gram-negative bacteria lose the stain. The Gram-negative bacteria subsequently stain with the safranin dye, the counterstain, used next. These bacteria appear red under the oil-immersion lens, while Gram-positive bacteria appear blue or purple, reflecting the crystal violet retained during the washing step.

Another differential stain technique is the **acid-fast technique.** This technique differentiates species of *Mycobacterium* from other bacteria. Heat or a lipid solvent is used to carry the first stain, carbolfuchsin, into the cells. Then the cells are washed with a dilute acid-alcohol solution. *Mycobacterium* species resist the effect of the acid-alcohol and retain the carbolfuchsin stain (bright red). Other bacteria lose the stain and take on the subsequent methylene blue stain (blue). Thus, the acid-fast bacteria appear bright red, while the nonacid-fast bacteria appear blue when observed under oil-immersion microscopy.

Other stain techniques seek to identify various bacterial structures of importance. For instance, a special stain technique highlights the **flagella** of bacteria by coating the flagella with dyes or metals to increase their width. Flagella so stained can then be observed.

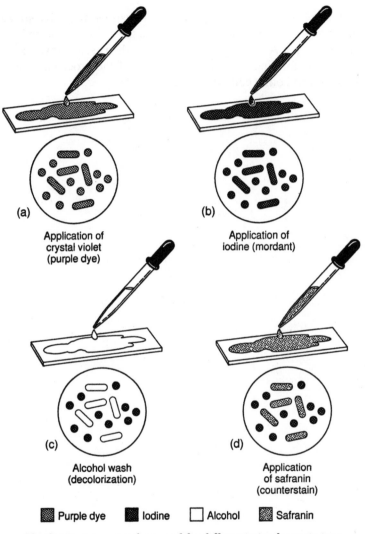

(a) Application of crystal violet (purple dye)

(b) Application of iodine (mordant)

(c) Alcohol wash (decolorization)

(d) Application of safranin (counterstain)

▨ Purple dye   ■ Iodine   □ Alcohol   ▨ Safranin

The Gram stain procedure used for differentiating bacteria into two groups.

■ Figure 6 ■

A special stain technique is used to examine bacterial **spores.** Malachite green is used with heat to force the stain into the cells and give them color. A counterstain, safranin, is then used to give color to the nonsporeforming bacteria. At the end of the procedure, spores stain green and other cells stain red.

Microorganisms and all other living organisms are classified as **prokaryotes** or **eukaryotes.** Prokaryotes and eukaryotes are distinguished on the basis of their cellular characteristics. For example, prokaryotic cells lack a nucleus and other membrane-bound structures known as organelles, while eukaryotic cells have both a nucleus and organelles (Figure 7).

Prokaryotic and eukaryotic cells are similar in several ways. Both types of cells are enclosed by cell membranes (plasma membranes), and both use DNA for their genetic information.

Prokaryotes include several kinds of microorganisms, such as bacteria and cyanobacteria. Eukaryotes include such microorganisms as fungi, protozoa, and simple algae. Viruses are considered neither prokaryotes nor eukaryotes because they lack the characteristics of living things, except the ability to replicate (which they accomplish only in living cells).

Prokaryotic Cells

The characteristics of **prokaryotic cells** apply to the bacteria and cyanobacteria (formerly known as blue-green algae), as well as to the rickettsiae, chlamydiae, and mycoplasmas.

Size and shape. Prokaryotes are probably the smallest living organisms, ranging in size from 0.15 μm (mycoplasmas) to 0.25 μm (chlamydiae) to 0.45 μm (rickettsiae) to about 2.0 μm (many of the bacteria). Certain prokaryotes, such as bacteria, occur in spherical forms called **cocci** (singular, **coccus**) or in rodlike forms called **bacilli** (singular, **bacillus**). Some bacteria have a comma shape (**vibrio**), or a flexible, wavy shape (**spirochete**), or a corkscrew shape (**spirillum**).

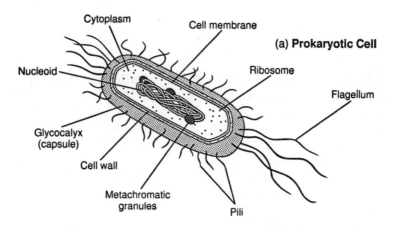

(a) Prokaryotic Cell

Cytoplasm

Cell membrane

Nucleoid

Ribosome

Flagellum

Glycocalyx
(capsule)

Cell wall

Metachromatic
granules

Pili

(b) Eukaryotic Cell

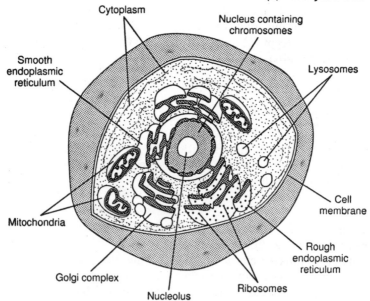

Cytoplasm

Nucleus containing
chromosomes

Smooth
endoplasmic
reticulum

Lysosomes

Cell
membrane

Mitochondria

Rough
endoplasmic
reticulum

Golgi complex

Nucleolus

Ribosomes

*The important cellular features of (a) a prokaryotic cell (a
bacterium) and (b) a eukaryotic cell.*

■ Figure 7 ■

Some prokaryotes have a variety of shapes and sizes and are said to be **pleomorphic.** Rickettsiae and mycoplasmas are examples of pleomorphic microorganisms.

When certain prokaryotes divide, they cling to each other in a distinct arrangement. A **diplococcus,** for example, consists of a pair of cocci, while a **streptococcus** consists of a chain of cocci, and a **tetracoccus** consists of four cocci arranged in a cube. A grapelike cluster of cocci is called a **staphylococcus.** Bacilli sometimes form long chains called **streptobacilli.**

The cell wall and cell membrane. With the exception of mycoplasmas, all bacteria have a semirigid **cell wall.** The cell wall gives shape to the organisms and prevents them from bursting, especially since materials in the cytoplasm exert osmotic pressures.

The chief component of the prokaryotic cell wall is **peptidoglycan,** a large polymer composed of N-acetylglucosamine and N-acetylmuramic acid. Gram-positive bacteria have more peptidoglycan in their cell wall, which may account for their ability to retain the stain in the Gram stain procedure. Gram-negative bacteria have more lipids in their cell wall. Polymers of **teichoic acid** are commonly associated with the peptidoglycan in Gram-positive bacteria.

In addition to the cell wall, Gram-negative bacteria have a very thin surrounding layer called the **outer membrane.** Lipopolysaccharides known as **endotoxins** are part of this outer membrane. A space called the **periplasmic space** separates the cell wall from the outer membrane and contains a substance called **periplasm.**

All prokaryotes have cytoplasm surrounded by a **cell membrane,** also known as the **plasma membrane.** The cell membrane conforms to the fluid mosaic model, which means that its proteins float within a double layer of phospholipids. Respiratory enzymes are located at the cell membrane of prokaryotes, and the membrane assists DNA replication and has attachment points for bacterial flagella.

The cytoplasm. The **cytoplasm** of prokaryotic cells contains ribosomes and various other granules used by the organism. The DNA is contained in the nuclear region (the **nucleoid**) and has no histone protein to support it. Prokaryotic cells have in their cytoplasm a single, looped **chromosome,** as well as numerous small loops of DNA called **plasmids.** Genetic information in the plasmids is apparently not essential for the continued survival of the organism.

Prokaryotic **ribosomes** contain protein and ribonucleic acid (RNA) and are the locations where protein is synthesized. Prokaryotic ribosomes have a sedimentation rate of 70S, and are therefore known as 70S ribosomes. (Eukaryotic cells have 80S ribosomes.) Certain antibiotics bind to these ribosomes and inhibit protein synthesis.

Some prokaryotic cells that engage in photosynthesis have internal membranes called **thylakoids** where their chlorophyll pigments are located. These membranes are also the sites of enzymes for photosynthesis. Certain bacteria have granules of phosphorus, starch, or glycogen. Granules called **metachromatic granules** stain with methylene blue and are used in diagnostic circumstances. Some bacterial species also have **magnetosomes,** which contain magnetic substances to help orient the organisms to hospitable environments.

External cellular structures. Many prokaryotic cells have at their surface a number of external structures that assist their functions. Among these structures are **flagella.** Flagella are found primarily in bacterial rods and are used for motility. A bacterium may have a single flagellum (a monotrichous bacterium), or flagella at both ends of the cell (an amphitrichous bacterium), or two or more flagella at one end of the cell (a lophotrichous bacterium), or it may be surrounded by flagella (a peritrichous bacterium).

Flagella are long, ultrathin structures, many times the length of the cell. They are composed of the protein **flagellin** arranged in long fibers. A hooklike structure and basal body connect the flagellum to the cell membrane. Flagella rotate and propel the bacteria.

Spirochetes lack flagella, but they possess **axial filaments.** The axial filaments extend beyond the cell wall and cause the spirochete to rotate in a corkscrew fashion and thereby move.

Some bacterial species have projections called **pili** (singular, **pilus**). Pili are used for attachments to surfaces such as tissues. Many pathogens possess pili, which are composed of the protein **pilin.** Certain pili, known as **conjugation pili,** unite prokaryotic cells to one another and permit the passage of DNA between the cells. The term **fimbriae** is often used for the attachment pili.

Many bacteria, especially pathogens, are enclosed at their surface by a layer of polysaccharides and proteins called the **glycocalyx.** The glycocalyx, composed of a thick, gummy material, serves as a reservoir for nutrients and protects the organism from changes in the environment. When the glycocalyx is a tightly bound structure, it is known as a **capsule.** When it is a poorly bound structure that flows easily, it is known as a **slime layer.** The material in dental plaque is composed largely of the material from the slime layer.

Endospores. Bacteria of the genera *Bacillus* and *Clostridium* are able to form highly resistant internal structures called **endospores,** or simply **spores.** Spores are formed during the normal life cycle when the environment becomes too harsh (Figure 8).

One vegetative (multiplying) cell produces one spore. Spores are able to withstand extremely high temperatures, long periods of drying, and other harsh environments. When conditions are favorable, the spore germinates and releases a new vegetative cell, which multiplies and reforms the colony. Sporeformers include the agents of anthrax, tetanus, botulism, and gas gangrene. Spores contain **dipicolinic acid** and calcium ions, both of which contribute to their resistance.

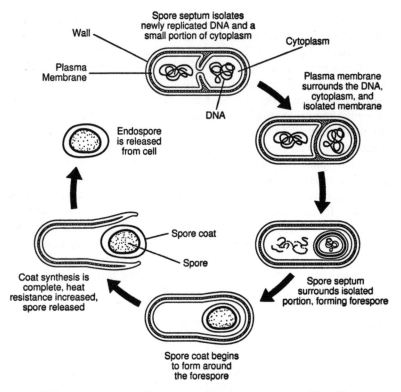

The process of spore formation as it occurs in species of Bacillus *and* Clostridium.

■ Figure 8 ■

Eukaryotic Cells

Eukaryotic cells are generally larger and more complex than prokaryotic cells. They also contain a variety of cellular bodies called organelles. The organelles function in the activities of the cell and are compartments for localizing metabolic function. Microscopic protozoa, unicellular algae, and fungi have eukaryotic cells.

Nucleus. Eukaryotic cells have a distinctive **nucleus,** composed primarily of protein and deoxyribonucleic acid, or DNA. The DNA is organized into linear units called **chromosomes,** also known as **chromatin** when the linear units are not obvious. Functional segments of the chromosomes are referred to as **genes.**

The nuclear proteins belong to a class of proteins called histones. **Histones** provide a supportive framework for the DNA in chromosomes. The DNA replicates in eukaryotic cells during the process of mitosis.

The nucleus of eukaryotic cells is surrounded by an outer membrane called the **nuclear envelope,** which is a double-membrane structure consisting of two lipid layers similar to the cell membrane. Pores exist in the nuclear membrane, and the internal nuclear environment can therefore communicate with the cytoplasm of the cell.

Within the nucleus are two or more dense masses referred to as **nucleoli** (singular, **nucleolus**). The nucleolus is an RNA-rich area where ribosomes are assembled before passing out of the nucleus into the cytoplasm.

Cellular organelles. Within the cytoplasm (also known as the cytosol) of eukaryotic cells are a number of microscopic bodies called **organelles** ("little organs"). Various functions of the cell go on within these organelles.

An example of an organelle is the **endoplasmic reticulum (ER),** a series of membranes that extend throughout the cytoplasm of eukaryotic cells. In some places the ER is studded with submicroscopic bodies called ribosomes. This type of ER is referred to as **rough ER.** In other places there are no ribosomes, and the ER is called **smooth ER.** The endoplasmic reticulum is the site of protein synthesis in the cell. Eukaryotic **ribosomes** are 80S bodies where the amino acids are bound together to form proteins. The spaces within the ER membranes are known as **cisternae.**

Another organelle is the **Golgi body** (also called the **Golgi apparatus**). The Golgi body is a series of flattened sacs, usually curled at the edges. The outermost sac often bulges away to form droplike vesicles known as **secretory vesicles.** It is in the Golgi body that the cell's proteins and lipids are processed and packaged before being sent to their final destination.

Another organelle, the **lysosome,** is derived from the Golgi body. It is a somewhat circular, droplike sac of enzymes within the cytoplasm. These enzymes are used for digestion in the cell. They break down the particles of food taken into the cell and make the products available to the cell. Enzymes are also contained in a cytoplasmic body called the **peroxisome.**

The organelle where much energy is released in the eukaryotic cell is the **mitochondrion** (plural, **mitochondria**). The energy released is used to form adenosine triphosphate (ATP). Because they are involved in energy release and storage, the mitochondria are called the "powerhouses of the cells."

An organelle found in certain protozoa is a large, fluid-filled, contractile **vacuole.** The vacuole may occupy over 75 percent of the cell interior and is used for eliminating water. Water pressure building up within the vacuole may cause the cell to swell.

Still another organelle within the cell is the **cytoskeleton,** an interconnected system of fibers, threads, and interwoven molecules that give structure to the cell. The main components of the cytoskeleton are **microtubules, microfilaments,** and **intermediate filaments.** All are assembled from subunits of protein.

Many eukaryotic cells contain flagella and cilia. Eukaryotic **flagella,** like prokaryotic flagella, are long, hairlike organelles that extend from the cell. Eukaryotic flagella whip about and propel the cell (as in protozoa) and are composed of nine pairs of microfilaments arranged about a central pair. **Cilia** are shorter and more numerous than flagella. In moving cells, they wave in synchrony and move the cell. *Paramecium* is a well-known ciliated protozoan.

The cell wall. Many species of eukaryotes, such as fungi, contain a **cell wall** outside the cell membrane. In fungi, the cell wall contains a complex polysaccharide called **chitin** as well as some cellulose. Algal cells, by contrast, have no chitin; rather, their cell walls are composed exclusively of the polysaccharide **cellulose.**

Cell walls provide support for eukaryotic cells and help the cells resist mechanical pressures while giving them a boxlike appearance. The cell walls are not selective devices, as are the cell membranes.

The cell membrane. The eukaryotic **cell membrane** conforms to the fluid mosaic model found in the prokaryotic membrane. In eukaryotes, the membrane is a dynamic structure governing passage of dissolved molecules and particles into and out from the cytoplasm. However, it neither contains the enzymes found in the prokaryotic cell nor functions in DNA replication.

In order for the cytoplasm of prokaryotic and eukaryotic cells to communicate with the external environment, materials must be able to move through the cell membrane. There are several mechanisms by which movement can occur. One method, called **diffusion,** is the movement of molecules from a region of high concentration to one of low concentration. This movement occurs because the molecules are constantly colliding with one another, and the net movement of the molecules is away from the region of high concentration. Diffusion is a random movement of molecules, and the pathway the molecules take is called the **concentration gradient.** Molecules are said to move down the concentration gradient in diffusion.

Another method of movement across the membrane is **osmosis,** the movement of water from a region of high concentration to one of low concentration. Osmosis occurs across a membrane that is **semipermeable,** meaning that the membrane lets only certain molecules pass through while keeping other molecules out. Osmosis is a type of diffusion involving only water.

A third mechanism for movement across the membrane is **facilitated diffusion,** a type of diffusion assisted by certain proteins in the membrane. The proteins permit only certain molecules to pass across the membrane and encourage movement from a region of high concentration of molecules to one of low concentration.

A fourth method for passing across the membrane is **active transport.** When active transport is taking place, a protein moves a certain material across the membrane from a region of low concentration to one of high concentration. Because this movement is happening against the concentration gradient, it requires that energy be expended, energy usually derived from ATP.

The final mechanism for movement across the cell membrane is **endocytosis,** a process in which a small patch of cell membrane encloses particles or tiny volumes of fluid at or near the cell surface. The membrane enclosure then sinks into the cytoplasm and pinches off from the membrane. When the vesicle contains particulate matter, the process is called **phagocytosis;** when it contains droplets of fluid, the process is called **pinocytosis.**

Microbial life can exist only where molecules and cells remain organized, and energy is needed by all microorganisms to maintain organization.

Every activity taking place in microbial cells involves both a shift of energy and a measurable loss of energy. Although the second law of thermodynamics says that energy cannot be created or destroyed, but only transferred within a system, unfortunately, the transfers of energy in living systems are never completely efficient. For this reason, considerably more energy must be taken into the system than is necessary to simply carry out the actions of microbial life.

Chemical Reactions and Energy

In microorganisms, most chemical compounds neither combine with one another automatically nor break apart automatically. A spark called the **energy of activation** is needed. The activation energy needed to spark an exergonic (energy-yielding) reaction or endergonic (energy-requiring) reaction can be heat energy or chemical energy. Reactions that require activation energy can also proceed in the presence of **biological catalysts.** Catalysts are substances that speed up chemical reactions but remain unchanged during the reactions. Catalysts work by lowering the required amount of activation energy for the chemical reaction. In microorganisms, the catalysts are enzymes.

Enzymes. Chemical reactions in microorganisms operate in the presence of **enzymes.** A particular enzyme catalyzes only one reaction, and thousands of different enzymes exist in a microbial cell to catalyze thousands of different chemical reactions. The substance acted on by an enzyme is called its **substrate.** The products of an enzyme-catalyzed chemical reaction are called **end products.**

All enzymes are composed of proteins. When an enzyme functions, a key portion of the enzyme called the **active site** interacts with the substrate. The active site closely matches the molecular configuration of the substrate, and after this interaction has taken place, a shape change at the active site places a physical stress on the substrate. This physical stress aids the alteration of the substrate and produces the end products. After the enzyme has performed its work, the product or products drift away. The enzyme is then free to function in the next chemical reaction. Enzyme-catalyzed reactions occur extremely fast.

With some exceptions, enzyme names end in "-ase." For example, the microbial enzyme that breaks down hydrogen peroxide to water and hydrogen is called catalase. Other well-known enzymes are amylase, hydrolase, peptidase, and kinase.

The rate of an enzyme-catalyzed reaction depends on a number of factors, including the concentration of the substrate, the acidity of the environment, the presence of other chemicals, and the temperature of the environment. For example, at higher temperatures, enzyme reactions occur more rapidly. Since enzymes are proteins, however, excessive amounts of heat may cause the protein to change its structure and become inactive. An enzyme altered by heat is said to be **denatured.**

Enzymes work together in metabolic pathways. A **metabolic pathway** is a sequence of chemical reactions occurring in a cell. A single enzyme-catalyzed reaction may be one of multiple reactions in the metabolic pathway. Metabolic pathways may be of two general types: Some involve the breakdown or digestion of large, complex molecules in the process of **catabolism.** Others involve a synthesis, generally by joining smaller molecules in the process of **anabolism.**

Many enzymes are assisted by chemical substances called **cofactors.** Cofactors may be ions or molecules associated with an enzyme and required in order for a chemical reaction to take place. Ions that might operate as cofactors include those of iron, manganese, or zinc. Organic molecules acting as cofactors are referred to as **coenzymes.** Examples of coenzymes are NAD and FAD (to be discussed shortly).

Adenosine triphosphate (ATP). Adenosine triphosphate (ATP) is the chemical substance that serves as the currency of energy in the microbial cell. It is referred to as currency because it can be "spent" in order to make chemical reactions occur.

ATP, used by virtually all microorganisms, is a nearly universal molecule of energy transfer. The energy released during the reactions of catabolism is stored in ATP molecules. In addition, the energy trapped in anabolic reactions such as photosynthesis is also trapped in ATP.

An ATP molecule consists of three parts (Figure 9a). One part is a double ring of carbon and nitrogen atoms called **adenine**. Attached to the adenine molecule is a small five-carbon carbohydrate called **ribose**. Attached to the ribose molecule are three **phosphate groups,** which are linked by covalent bonds.

The covalent bonds that unite the phosphate units in ATP are high-energy bonds. When an ATP molecule is broken down by an enzyme, the third (terminal) phosphate unit is released as a phosphate group, which is a phosphate ion (Figure 9b). With the release,

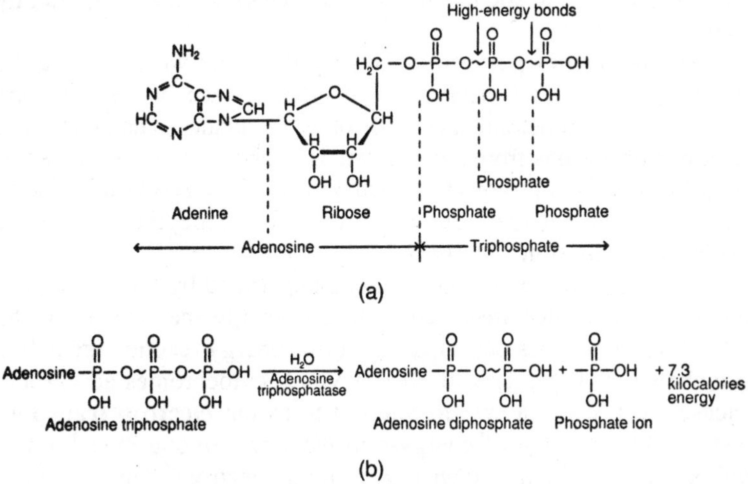

(a)

(b)

The adenosine triphosphate (ATP) molecule that serves as an immediate energy source in the cell.

■ Figure 9 ■

approximately 7.3 kilocalories of energy (a kilocalorie is 1000 calories) are made available to do the work of the microorganism.

The breakdown of an ATP molecule is accomplished by an enzyme called adenosine triphosphatase. The products of ATP breakdown are **adenosine diphosphate (ADP)** and, as noted, a **phosphate ion.** Adenosine diphosphate and the phosphate ion can be reconstituted to form ATP, much as a battery can be recharged. To accomplish this ATP formation, energy necessary for the synthesis can be made available in the microorganism through two extremely important processes: photosynthesis and cellular respiration. A process called fermentation may also be involved.

ATP production. ATP is generated from ADP and phosphate ions by a complex set of processes occurring in the cell, processes that depend upon the activities of a special group of cofactors called coenzymes. Three important coenzymes are nicotinamide adenine dinucleotide (**NAD**), nicotinamide adenine dinucleotide phosphate (**NADP**), and flavin adenine dinucleotide (**FAD**). All are structurally similar to ATP.

All **coenzymes** perform essentially the same work. During the chemical reactions of metabolism, coenzymes accept electrons and pass them on to other coenzymes or other molecules. The removal of electrons or protons from a coenzyme is called **oxidation.** The addition of electrons or protons to a coenzyme is called **reduction.** Therefore, the chemical reactions performed by coenzymes are called **oxidation-reduction reactions.**

The oxidation-reduction reactions performed by the coenzymes and other molecules are essential to the energy metabolism of the cell. Other molecules participating in this energy reaction are called **cytochromes.** Together with the enzymes, cytochromes accept and release electrons in a system referred to as the **electron transport system.** The passage of energy-rich electrons among cytochromes and coenzymes drains the energy from the electrons. This is the energy used to form ATP from ADP and phosphate ions.

The actual formation of ATP molecules requires a complex process referred to as **chemiosmosis.** Chemiosmosis involves the creation of a steep proton gradient, which occurs between the membrane-bound areas. In prokaryotic cells (for example, bacteria), it is the area of the cell membrane; in eukaryotic cells, it is the membranes of the mitochondria. A gradient is formed when large numbers of protons (hydrogen ions) are pumped into membrane-bound compartments. The protons build up dramatically within the compartment, finally reaching an enormous number. The energy used to pump the protons is energy released from the electrons during the electron transport system.

After large numbers of protons have gathered at one side of the membrane, they suddenly reverse their directions and move back across the membranes. The protons release their energy in this motion, and the energy is used by enzymes to unite ADP with phosphate ions to form ATP. The energy is trapped in the high-energy bond of ATP by this process, and the ATP molecules are made available to perform cell work.

Cellular Respiration

Microorganisms such as cyanobacteria can trap the energy in sunlight through the process of photosynthesis and store it in the chemical bonds of carbohydrate molecules. The principal carbohydrate formed in photosynthesis is glucose. Other types of microorganisms such as nonphotosynthetic bacteria, fungi, and protozoa are unable to perform this process. Therefore, these organisms must rely upon preformed carbohydrates in the environment to obtain the energy necessary for their metabolic processes.

Cellular respiration is the process by which microorganisms obtain the energy available in carbohydrates. They take the carbohydrates into their cytoplasm, and through a complex series of metabolic processes, they break down the carbohydrate and release the energy. The energy is generally not needed immediately, so it is used to combine ADP with phosphate ions to form ATP molecules. During the process of cellular respiration, **carbon dioxide** is given off as a waste product. This carbon dioxide can be used by photosynthesizing cells to form new carbohydrates. Also in the process of cellular respiration, oxygen gas is required to serve as an acceptor of electrons. This oxygen gas is identical to the oxygen gas given off in photosynthesis.

The overall mechanism of cellular respiration involves four subdivisions: **glycolysis,** in which glucose molecules are broken down to form pyruvic acid molecules; the **Krebs cycle,** in which pyruvic acid is further broken down and the energy in its molecule is used to form high-energy compounds such as NADH; the **electron transport system,** in which electrons are transported along a series of coenzymes and cytochromes and the energy in the electrons is released; and **chemiosmosis,** in which the energy given off by electrons is used to pump protons across a membrane and provide the energy for ATP synthesis.

Glycolysis. The process of **glycolysis** is a multistep metabolic pathway that occurs in the cytoplasm of microbial cells and the cells of other organisms. At least six enzymes operate in the metabolic pathway.

In the first and third steps of the pathway, ATP is used to energize the molecules. Thus, two molecules of ATP must be expended in the process. Further along in the process, the six-carbon glucose molecule is converted into intermediary compounds and then is split into two three-carbon compounds. The latter undergo additional conversions and eventually form **pyruvic acid** at the conclusion of the process.

During the latter stages of glycolysis, four ATP molecules are synthesized using the energy given off during the chemical reactions. Thus, four ATP molecules are synthesized and two ATP molecules are inserted into the process for a net gain of two ATP molecules in glycolysis.

Also during glycolysis, another of the reactions yields enough energy to convert NAD to **NADH.** The reduced coenzyme (NADH) will later be used in the electron transport system, and its energy will be released. During glycolysis, two NADH molecules are produced.

As glycolysis does not use oxygen, the process is considered to

be anaerobic. For certain anaerobic organisms, such as certain bacteria and fermentation yeasts, glycolysis is the sole source of energy. It is a somewhat inefficient process because much of the cellular energy remains in the two molecules of pyruvic acid.

The Krebs cycle. Following glycolysis, the mechanism of cellular respiration then involves another multistep process called the **Krebs cycle,** also called the citric acid cycle and the tricarboxylic acid cycle. The Krebs cycle uses the two molecules of pyruvic acid formed in glycolysis and yields high-energy molecules of NADH and FADH and some ATP and carbon dioxide (Figure 10).

The Krebs cycle occurs at the cell membrane of bacterial cells and in the **mitochondria** of eukaryotic cells. Each of these sausage-shaped organelles of eukaryotic microorganisms possesses inner and outer membranes, and therefore an inner and outer compartment. The inner membrane is folded over itself many times; the folds are called **cristae.** Along the cristae are the important enzymes necessary for the proton pump and for ATP production.

Prior to entering the Krebs cycle, the pyruvic acid molecules are processed. Each three-carbon molecule of pyruvic acid undergoes conversion to a substance called acetyl-coenzyme A, or **acetyl-CoA.** In the process, the pyruvic acid molecule is broken down by an enzyme, one carbon atom is released in the form of carbon dioxide, and the remaining two carbon atoms are combined with a coenzyme called coenzyme A. This combination forms acetyl-CoA. In the process, electrons and a hydrogen ion are transferred to NAD to form high-energy **NADH.**

Acetyl-CoA now enters the Krebs cycle by combining with a four-carbon acid called oxaloacetic acid. The combination forms the six-carbon acid called **citric acid.** Citric acid undergoes a series of enzyme-catalyzed conversions. The conversions, which involve up to 10 chemical reactions, are all brought about by enzymes. In many of the steps, high-energy electrons are released to NAD. The NAD molecule also acquires a hydrogen ion and becomes NADH. In one of the steps, FAD serves as the electron acceptor, and it acquires two hydrogen ions to become $FADH_2$. Also, in one of the reactions, enough

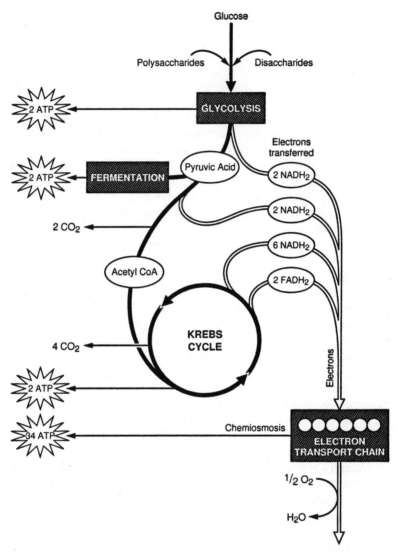

*An overview of the processes of cellular respiration showing the
major pathways and the places where ATP is synthesized.*

■ Figure 10 ■

energy is released to synthesize a molecule of ATP. Since there are two pyruvic acid molecules entering the system, two ATP molecules are formed.

Also during the Krebs cycle, the two carbon atoms of acetyl-CoA are released and each forms a carbon dioxide molecule. Thus, for each acetyl-CoA entering the cycle, two carbon dioxide molecules are formed. Since two acetyl-CoA molecules enter the cycle, and each has two carbon atoms, four carbon dioxide molecules will form. Add these four molecules to the two carbon dioxide molecules formed in the conversion of pyruvic acid to acetyl-CoA, and the total is six carbon dioxide molecules. These six CO_2 molecules are given off as waste gas in the Krebs cycle. They represent the six carbons of glucose that originally entered the process of glycolysis.

At the end of the Krebs cycle, the final product formed is **oxalo-acetic acid,** identical to the oxaloacetic acid which begins the cycle. The molecule is now ready to accept another acetyl-CoA molecule to begin another turn of the cycle. All told, the Krebs cycle forms (per two molecules of pyruvic acid) two ATP molecules, a large number of NADH molecules, and some $FADH_2$ molecules. The NADH and the $FADH_2$ will be used in the electron transport system.

The electron transport system. The **electron transport system** occurs at the bacterial cell membrane and in the cristae of the mitochondria in eukaryotic cells. Here, a series of **cytochromes** (cell pigments) and **coenzymes** exist. These cytochromes and coenzymes act as carrier molecules and transfer molecules. They accept high-energy electrons and pass the electrons to the next molecule in the system. At key proton-pumping sites, the energy of the electrons is used to transport protons across the cell membrane or into the outer compartment of the mitochondrion.

Each NADH molecule is highly energetic. It accounts for the transfer of six protons across the membrane. Each $FADH_2$ molecule accounts for the transfer of four protons. Electrons pass from NAD to FAD, to other cytochromes and coenzymes, and eventually they lose much of their energy. The final electron acceptor is an **oxygen** atom. The electron-oxygen combination then takes on two protons to form a molecule of **water** (H_2O). As a final electron receptor, oxygen is responsible for removing electrons from the system. If oxygen were not available, electrons could not be passed among the coenzymes, the energy in electrons could not be released, the proton pump could not be established, and ATP could not be produced.

Chemiosmosis. The actual production of ATP in cellular respiration takes place during **chemiosmosis.** As previously noted, chemiosmosis involves the pumping of protons through special channels in the membranes of mitochondria from the inner to the outer compartment. In bacteria, the punping occurs at the cell membrane. The pumping establishes a proton gradient. Once the gradient is established, protons pass down the gradient through molecular particles. In these particles, the energy of the protons is used to generate ATP, using ADP and phosphate ions as the starting points.

The energy production in cellular respiration during chemiosmosis is substantial. Most biochemists agree that in prokaryotic microorganisms, a total of 36 molecules of ATP can be produced during cellular respiration. In eukaryotic cells, the number is 34 molecules of ATP. Two molecules of ATP are produced as the net gain of glycolysis, so the grand total is 38 molecules of ATP (36 in eukaryotes). These ATP molecules may then be used in the cell for its needs.

Fermentation. **Fermentation** is an anaerobic process in which energy can be released from glucose even though oxygen is not available. Fermentation occurs in yeast cells, and a form of fermentation takes place in bacteria.

In **yeast cells,** glucose can be metabolized through cellular respiration, as in other cells. When oxygen is lacking, however, glucose is still changed to pyruvic acid via glycolysis. The pyruvic acid is first converted to acetaldehyde and then to **ethyl alcohol.** The net gain of ATP to the yeast cell is two molecules—the two molecules of ATP normally produced in glycolysis.

Yeasts are able to participate in fermentation because they have the necessary enzyme to convert pyruvic acid to ethyl alcohol. This process is essential because it removes electrons and hydrogen ions from NADH during glycolysis. The effect is to free the NAD so that it can participate in future reactions of glycolysis. Yeasts are therefore used in both bread and alcohol production. Alcohol fermentation is the process that yields beer, wine, and other spirits. The carbon dioxide given off supplements the carbon dioxide given off during the Krebs cycle and causes bread to rise.

Photosynthesis

A great variety of living things on earth, including all photosynthetic microorganisms, synthesize their foods from simple molecules such as carbon dioxide and water. In microorganisms, photosynthesis occurs in unicellular algae and in photosynthesizing bacteria such as cyanobacteria and green and purple sulfur bacteria.

Photosynthesis is actually two processes. In the first process, energy-rich electrons flow through a series of coenzymes and other molecules, and this electron energy is trapped. During the trapping process, ATP molecules and molecules of **nicotinamide adenine dinucleotide phosphate hydrogen (NADPH)** are formed, both rich in energy. These molecules are then used in the second half of the process, where carbon dioxide molecules are bound into carbohydrates to form organic substances such as glucose.

Photosynthesis occurs along the **thylakoid** membranes of eukaryotic organisms. The thylakoids are somewhat similar to the cristae of mitochondria. Sunlight is captured in the thylakoid by pigment molecules organized into photosystems. The coenzyme NADP functions in the system. The **photosystem** includes the pigment molecules, as well as proton pumps, and coenzymes and molecules of electron transport systems. In prokaryotic microorganisms, the chlorophyll molecules are dissolved in the cell's cytoplasm and are called bacteriochlorophylls.

The process of photosynthesis is conveniently divided into two parts: the energy-fixing reaction (also called the light reaction) and the carbon-fixing reaction (also called the light-independent reaction, or the dark reaction).

The energy-fixing reaction. The **energy-fixing reaction** of photosynthesis begins when light is absorbed in a photosystem. The energy activates electrons to jump out of chlorophyll molecules in the reaction center. These electrons pass through a series of cytochromes in the nearby electron transport system. Some of the energy of the electrons is lost as they move along the chain of acceptors, but a portion of the energy is used to pump protons across a membrane, setting up the potential for chemiosmosis.

After passing through the electron transport system, the energy-rich electrons enter another photosystem. Light now activates the electrons, and they receive a second boost out of the chlorophyll molecules. The electrons progress through a second electron transport system and enter a molecule of NADP. Since NADP has acquired two negatively charged electrons, it attracts two positively charged protons from a water molecule to balance the charges, and the molecule is reduced to NADPH. This molecule contains much energy.

Because electrons have flowed out of the chlorophyll molecules, the latter are left without a certain number of electrons. These electrons are replaced by electrons secured from water molecules. The third product of the disrupted water molecules is oxygen. Two oxygen atoms combine with one another to form molecular oxygen. This oxygen is given off by cyanobacteria as the waste product of photosynthesis. It is the oxygen that fills the atmosphere and is used by all oxygen-breathing organisms.

What has been described above are the **noncyclic energy-fixing reactions.** Certain microorganisms are also known to participate in **cyclic energy-fixing reactions.** Excited electrons leave the chlorophyll molecules, pass through coenzymes of the electron transport system, and then follow a special pathway back to the chlorophyll molecules. Each electron powers the proton pump and encourages the transport of a proton across the membrane. This process enriches the proton gradient and eventually leads to the generation of ATP.

ATP production in the energy-fixing reactions of photosynthesis occurs by the process of **chemiosmosis.** Essentially, it consists of a rush of proteins across a membrane (the microbial membrane, in this case) accompanied by the synthesis of ATP molecules. It has been calculated that the proton concentration on one side of the membrane is 10,000 times that on the opposite side of the membrane.

In photosynthesis, the protons pass back across the membranes through channels that lie alongside sites where enzymes are located. As the protons pass through the channels, the energy of the protons is released to form high-energy ATP bonds. ATP is formed in the energy-fixing reactions along with NADPH formed in the main reactions. Both ATP and NADPH provide the energy necessary for the synthesis of carbohydrates that occurs in the second major set of events in photosynthesis.

The carbon-fixing reaction. In the **carbon-fixing reaction** of photosynthesis, glucose and other carbohydrates are synthesized. This phase of photosynthesis occurs in the cytoplasm of the microbial cell.

In the carbon-fixing stage, an essential material, **carbon dioxide,** is obtained from the atmosphere. The carbon dioxide is attached to a five-carbon compound called **ribulose biphosphate (RuBP)** to form a six-carbon product. This product immediately breaks into two three-carbon molecules (Figure 11).

The three-carbon molecule is called **phosphoglycerate (PGA).** Each phosphoglycerate molecule is converted to **phosphoglyceraldehyde (PGAL)** using the ATP and NADPH synthesized in the energy-fixing reaction. The organic compounds that result have three carbon atoms. They interact with one another and eventually join to form a single molecule of six-carbon **glucose.** The process also generates more molecules of ribulose biphosphate to participate in further carbon-fixing reactions.

The carbon-fixing reaction is often referred to as the Calvin cycle, for Melvin Calvin, who performed much of the biochemical research. The product of the reaction is glucose, a carbohydrate containing the energy of sunlight, which began the reactions in the chlorophyll molecule. This energy has passed through ATP and NADPH and is now present in the high-energy glucose molecules. Photosynthesizing microorganisms use the glucose to obtain the energy for their activities. Nonphotosynthesizing organisms use this same glucose by consuming the carbohydrate.

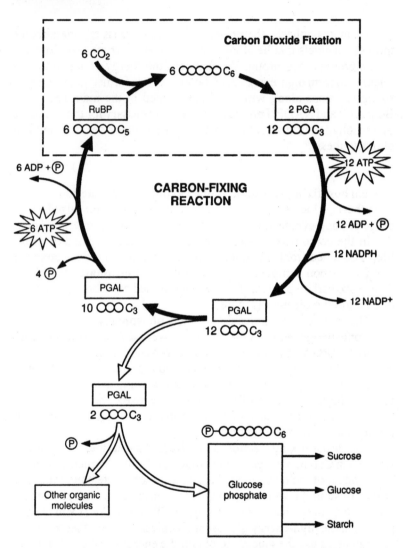

The carbon-fixing reaction of photosynthesis. The ATP and NADPH used in this process are synthesized in the energy-fixing phase.

■ Figure 11 ■

A characteristic of microorganisms is their ability to grow and form a population of organisms. One of the results of microbial metabolism is an increase in the size of the cell. The many requirements for successful growth include those both chemical and physical.

Growth Requirements

Chemical requirements. In order to grow successfully, microorganisms must have a supply of water as well as numerous other substances including mineral elements, growth factors, and gas, such as oxygen. Virtually all chemical substances in microorganisms contain **carbon** in some form, whether they be proteins, fats, carbohydrates, or lipids. Perhaps 50 percent of a bacterium's dry weight is carbon. Carbon can be obtained from organic materials in the environment, or it may be derived from carbon dioxide. Both chemoautotrophic and photoautotrophic microorganisms obtain their energy and produce their nutrients from simple inorganic compounds such as carbon dioxide. **Chemoautotrophs** do so through chemical reactions, while **photoautotrophs** use photosynthesis.

Among the other elements required by microorganisms are nitrogen and phosphorous. **Nitrogen** is used for the synthesis of proteins, amino acids, DNA, and RNA. Bacteria that obtain nitrogen directly from the atmosphere are called nitrogen-fixing bacteria. They include species of *Rhizobium* and *Azotobacter*, both found in the soil. **Phosphorus** is an essential element for nucleic acid synthesis and for the construction of phospholipids.

Oxygen is used by aerobic bacteria during the process of cellular respiration as a final electron acceptor. For **aerobic** organisms, oxygen is an absolute requirement for their energy-yielding properties. Certain microorganisms grow in oxygen-free environments and are described as **anaerobic.** Organisms such as these produce odoriferous gases in their metabolism, including hydrogen sulfide gas and methane. Certain pathogenic species, such as *Clostridium* species, are anaerobic.

Certain species of microorganisms are said to be **facultative.** These species grow in either the presence or absence of oxygen. Some bacteria species are **microaerophilic,** meaning that they grow in low concentrations of oxygen. In some cases, these organisms must have an environment rich in carbon dioxide. Organisms such as these are said to be **capnophilic.**

Other chemical requirements for microbial growth include such **trace elements** as iron, copper, and zinc. These elements often are used for the synthesis of enzymes. Organic growth factors such as vitamins may also be required by certain bacteria. Amino acids, purines, and pyrimidines should also be available.

Physical requirements. Certain physical conditions affect the type and amount of microbial growth. For example, enzyme activity depends on the **temperature** of the environment, and microorganisms are classified in three groups according to their temperature preferences: **psychrophilic** organisms (psychrophiles) prefer cold temperatures of about 0°C to 20°C; **mesophilic** organisms (mesophiles) prefer temperatures at 20°C to 40°C; **thermophilic** organisms (thermophiles) prefer temperatures higher than 40°C (Figure 12). A minimum and a maximum growth temperature range exist for each species. The temperature at which best growth occurs is the **optimum growth temperature.**

Another physical requirement is the extent of acidity or alkalinity, referred to as the **pH** of a solution. For most bacteria, the optimum pH is between 6.5 and 7.5. Since the pH of most human tissue is 7.0 to 7.2, these **neutrophilic** bacteria usually grow well in the body. Certain bacteria, such as those in sauerkraut and yogurt, prefer acidic environments of 6.0 or below. These bacteria are said to be **acidophilic.** Molds and yeasts are among other common acidophilic microorganisms.

Microbial growth proceeds best when the **osmotic pressure** is ideal. Normally, the salt concentration of microbial cytoplasm is about 1 percent. When the external environment also has a 1 percent salt

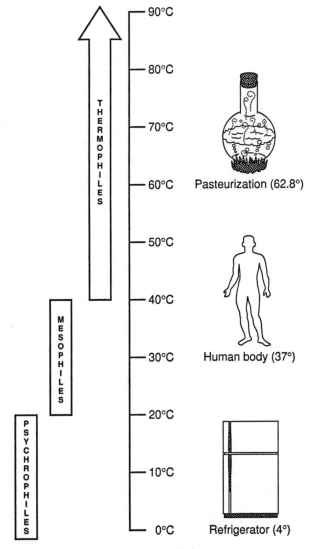

Three types of bacteria and the temperature environments in which they thrive.

■ Figure 12 ■

concentration, then the osmotic pressure is optimum. Should the external salt concentration rise, as when food is salted, water will flow out of the microbial cytoplasm by osmosis through the cell membrane into the environment, thereby causing the microorganisms to shrink and die. By comparison, if exterior water is free of salt, it will flow through the cell membrane into the cytoplasm of the cell, causing the organism to swell and burst.

Microorganisms that live in marine environments can tolerate high salt concentrations. These organisms are said to be **halophilic.** They include diatoms and dinoflagellates, two types of unicellular algae that lie at the base of oceanic food chains. There are many other species of halophilic bacteria, fungi, protozoa, and algae.

Microbial Cultivation

When microorganisms are cultivated in the laboratory, a growth environment called a **medium** is used. The medium may be purely chemical (a chemically defined medium), or it may contain organic materials, or it may consist of living organisms such as fertilized eggs. Microorganisms growing in or on such a medium form a **culture.** A culture is considered a **pure culture** if only one type of organism is present and a **mixed culture** if populations of different organisms are present. When first used, the culture medium should be sterile, meaning that no form of life is present before inoculation with the microorganism.

General microbial media. For the cultivation of bacteria, a commonly used medium is **nutrient broth,** a liquid containing proteins, salts, and growth enhancers that will support many bacteria. To solidify the medium, an agent such as **agar** is added. Agar is a polysaccharide that adds no nutrients to a medium, but merely solidifies it. The medium that results is **nutrient agar.**

Many media for microorganisms are complex, reflecting the growth requirements of the microorganisms. For instance, most fungi require extra carbohydrate and an acidic environment for optimal growth. The medium employed for these organisms is **potato dextrose agar,** also known as **Sabouraud dextrose agar.** For protozoa, liquid media are generally required, and for rickettsiae and viruses, living tissue cells must be provided for best cultivation.

For anaerobic microorganisms, the atmosphere must be oxygen free. To eliminate the oxygen, the culture media can be placed within containers where carbon dioxide and hydrogen gas are generated and oxygen is removed from the atmosphere. Commercially available products achieve these conditions. Anaerobic chambers can also be used within closed compartments, and technicians can manipulate culture media within these chambers. To encourage carbon dioxide formation, a candle can be burned to use up oxygen and replace it with carbon dioxide.

Special microbial media. Certain microorganisms are cultivated in **selective media.** These media retard the growth of unwanted organisms while encouraging the growth of the organisms desired. For example, **mannitol salt agar** is selective for staphylococci because most other bacteria cannot grow in its high-salt environment. Another selective medium is **brilliant green agar,** a medium that inhibits Gram-positive bacteria while permitting Gram-negative organisms such as *Salmonella* species to grow.

Still other culture media are **differential media.** These media provide environments in which different bacteria can be distinguished from one another. For instance, **violet red bile agar** is used to distinguish coliform bacteria such as *Escherichia coli* from noncoliform organisms. The coliform bacteria appear as bright pink colonies in this media, while noncoliforms appear a light pink or clear.

Certain media are both selective and differential. For instance, **MacConkey agar** differentiates lactose-fermenting bacteria from nonlactose-fermenting bacteria while inhibiting the growth of Gram-positive bacteria. Since lactose-fermenting bacteria are often involved in water pollution, they can be distinguished by adding samples of water to MacConkey agar and waiting for growth to appear.

In some cases, it is necessary to formulate an **enriched medium.** Such a medium provides specific nutrients that encourage selected species of microorganisms to flourish in a mixed sample. When attempting to isolate *Salmonella* species from fecal samples, for instance, it is helpful to place a sample of the material in an enriched medium to encourage *Salmonella* species to multiply before the isolation techniques begin.

In order to work with microorganisms in the laboratory, it is desirable to obtain them in pure cultures. Pure cultures of bacteria can be obtained by spreading bacteria out and permitting the individual cells to form masses of growth called **colonies.** One can then pick a sample from the colony and be assured that it contains only one kind of bacteria. Cultivating these bacteria on a separate medium will yield a pure culture.

To preserve microbial cultures, they may be placed in the refrigerator to slow down the metabolism taking place. Two other methods are deep-freezing and freeze-drying. For deep-freezing, the microorganisms are placed in a liquid and frozen quickly at temperatures below −50°C. Freeze-drying (lyophilization) is performed in an apparatus that uses a vacuum to draw water off after the microbial suspension has been frozen. The culture resembles a powder, and the microorganisms can be preserved for long periods in this condition.

Isolation methods. To obtain separated colonies from a mixed culture, various **isolation methods** can be used. One is the **streak plate method,** in which a sample of mixed bacteria is streaked several times along one edge of a Petri dish containing a medium such as nutrient agar. A loop is flamed and then touched to the first area to retrieve a sample of bacteria. This sample is then streaked several times in the second area of the medium. The loop is then reflamed, touched to the second area, and streaked once again in the third area. The process can be repeated in a fourth and fifth area if desired. During incubation, the bacteria will multiply rapidly and form colonies (Figure 13a).

A second isolation method is the **pour plate method.** In this method, a sample of bacteria is diluted in several tubes of melted medium such as nutrient agar. After dilution, the melted agar is poured into separate Petri dishes and allowed to harden. Since the bacteria have been diluted in the various tubes, the plates will contain various dilutions of bacteria, and where the bacteria are most diluted, they will form isolated colonies (Figure 13b).

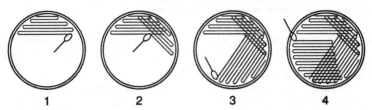

(a) Steps in the Streak Plate Method

1   2   3   4

(b) Steps in the Pour Plate Method

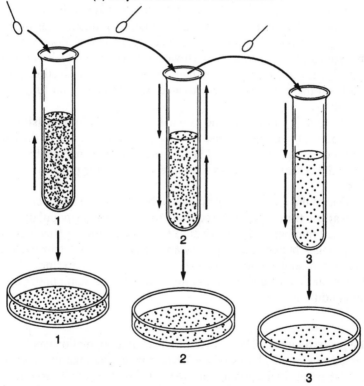

1   2   3

*Two processes for isolating bacteria from a mixed culture.
(a) The streak plate technique. (b) The pour plate technique.*

■ Figure 13 ■

Microbial Reproduction and Growth

Reproduction patterns. During their growth cycles, microorganisms undergo reproduction many times, causing the numbers in the population to increase dramatically.

In fungi, unicellular algae, and protozoa, **reproduction** involves a duplication of the nucleus through the asexual process of mitosis and a splitting of the cell in cytokinesis. Reproduction can also occur by a sexual process in which haploid nuclei unite to form a diploid cell having two sets of chromosomes. Various changes then follow to yield a sexually produced offspring. Sexual reproduction has the advantage of mixing chromosomes to obtain genetic variations not possible with asexual reproduction. However, fewer individuals normally result from sexual reproduction than from asexual reproduction. More details on these methods are provided in the chapters on fungi and protozoa.

Bacteria reproduce by the asexual process of **binary fission.** In this process, the chromosomal DNA duplicates, after which the bacterial membrane and cell wall grow inward to meet one another and divide the cell in two. The two cells separate and the process is complete.

One of the remarkable attributes of bacteria is the relatively short **generation time,** the time required for a microbial population to double in numbers. The generation time varies among bacteria and often ranges between 30 minutes and three hours. Certain bacteria have very brief generation times. *Escherichia coli*, for example, has a generation time of about 20 minutes when it is dividing under optimal conditions.

The growth curve. The growth of a bacterial population can be expressed in various phases of a **growth curve.** The logarithms of the actual numbers in the population are plotted in the growth curve along the side axis, and the time is plotted at the base. Four phases of growth are recognized in the growth curve.

In the first phase, called the **lag phase,** the population remains at the same number as the bacteria become accustomed to their new environment. Metabolic activity is taking place, and new cells are being produced to offset those that are dying.

In the **logarithmic phase,** or **log phase,** bacterial growth occurs at its optimal level and the population doubles rapidly. This phase is represented by a straight line, and the population is at its metabolic peak. Research experiments are often performed at this time.

During the next phase, the **stationary phase,** the reproduction of bacterial cells is offset by their death, and the population reaches a plateau. The reasons for bacterial death include the accumulation of waste, the lack of nutrients, and the unfavorable environmental conditions that may have developed. If the conditions are not altered, the population will enter its **decline,** or **death phase** (Figure 14). The bacteria die off rapidly, the curve turns downward, and the last cell in the population soon dies.

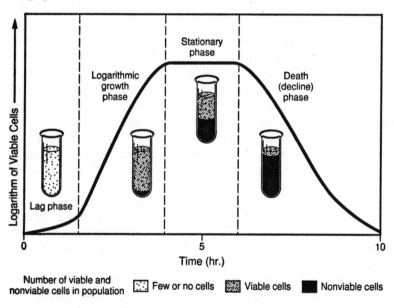

A growth curve of a bacterial population showing the four major phases of the curve.

■ Figure 14 ■

Microbial measurements. In order to measure the number of bacteria in a population, various methods are available. In one method, known as the **plate count method,** a sample of bacteria is diluted in saline solution, distilled water, or other holding fluid. Samples of the dilutions are then placed in Petri dishes with a growth medium and set aside to incubate. Following incubation, the count of colonies is taken and multiplied by the dilution factor represented by that plate. Generally, plates with between 30 and 300 colonies are selected for determining the final count, which is expressed as the number of bacteria per original ml of sample.

Another measuring method is to determine the **most probable number.** This technique is often used to determine the number of bacteria in a sample of contaminated water. Water samples are added to numerous tubes of single-strength and double-strength lactose broth. If coliform bacteria (such as *E. coli*) are present, they will ferment the lactose and produce gas. Judging by the number of tubes that contain gas at the end of the test, one may approximate the original number of bacteria in the water sample.

Another evaluative method is by a **direct microscopic count.** A specially designed counting chamber called a Petroff-Hausser counter is used. A measured sample of the bacterial suspension is placed on the counter, and the actual number of organisms is counted in one section of the chamber. Multiplying by an established reference figure gives a number of bacteria in the entire chamber and in the sample counted. The disadvantage of this method is that both live and dead bacteria are counted.

Turbidity methods can also be used to assess bacterial growth. As bacteria multiply in liquid media, they make the media cloudy. Placing the culture tube in a beam of light and noting the amount of light transmitted gives an idea of the turbidity of the culture and the relative number of bacteria it contains.

The **dry weight** of a culture can also be used to determine microbial numbers. The liquid culture is dried out, and the amount of microbial mass is weighed on a scale. It is also possible to measure the **oxygen uptake** of a culture of bacteria. If more oxygen is used by culture A than by culture B and all other things are equal, then it may be deduced that more microorganisms are present in culture A. A variation of this method called the **biochemical oxygen demand (BOD)** is used to measure the extent of contamination in a water sample.

The control of microbial growth may involve sterilization, disinfection, antisepsis, sanitization, or degerming. **Sterilization** is the destruction of all forms of microbial life, with particular attention to bacterial spores. Disinfection and antisepsis both refer to destruction of microbial pathogens, although some organisms, such as bacterial spores, may remain alive. **Disinfection** refers to the destruction of pathogenic organisms on an inanimate (lifeless) object, such as a tabletop, while **antisepsis** refers to that destruction on a living object, such as the skin surface.

Sanitization refers to the reduction in the number of pathogens to a level deemed safe by public health guidelines. **Degerming** is the physical removal of microorganisms by using such things as soaps or detergents.

Any chemical agent that kills microorganisms is known as a **germicide.** An agent that destroys bacteria is called a **bactericide,** one that kills fungi is a **fungicide,** and one that kills viruses is a **viricide.** A **bacteriostatic agent** prevents the further multiplication of bacteria without necessarily killing all that are present.

Among the conditions affecting the use of a germicide are temperature, the type of microorganism, and the environment. Germicides are more effective at high temperatures because the chemical breaks down at lower temperatures. Microorganisms vary in their susceptibility depending on such things as the composition of their cell wall, the presence or absence of a capsule, and the ability to form spores or cysts. The environment can affect the activity of a germicide, as, for example, when organic matter is present. This material shields microorganisms from germicides and often reacts with the germicide.

Physical Methods of Control

Physical methods for controlling the growth of microorganisms can be divided into heat methods and nonheat methods. The lowest temperature at which all microorganisms are killed in 10 minutes is the **thermal death point**, while the minimum amount of time required to kill microorganisms at a given temperature is known as the **thermal death time.** The time for destruction of 90 percent of the microbial population is the **decimal reduction time.**

Dry heat. Dry heat kills microorganisms by reacting with and oxidizing their proteins. Dry heat can be used in incineration devices, such as the **Bunsen burner** or the **hot-air oven.** In the hot-air oven, a temperature of about 170°C for two hours will bring about sterilization.

Moist heat. Moist heat is used to kill microorganisms in such things as **boiling water.** Most vegetating microorganisms are killed within two or three minutes, but over two or three hours may be required for destruction of bacterial spores. In moist heat, the microbial proteins undergo **denaturation,** a process in which the three-dimensional form of the protein reverts to a two-dimensional form, and the protein breaks down.

Moist heat is used in the **autoclave,** a high-pressure device in which steam is superheated (Figure 15). Steam at 100°C is placed under a pressure of 15 pounds per square inch, increasing the temperature to 121°C. At this temperature, the time required to achieve sterilization is about 15 minutes. The autoclave is the standard instrument for preparing microbial media and for sterilizing instruments such as syringes, hospital garb, blankets, intravenous solutions, and myriad other items.

Although **pasteurization** is used to lower the bacterial content of milk and dairy products, it does not achieve sterilization. The conditions of pasteurization are set up to eliminate the tuberculosis bacillus

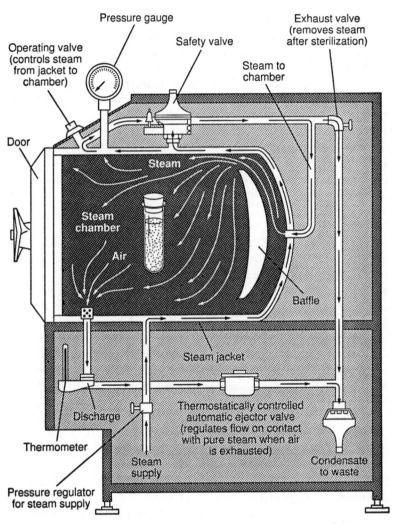

The autoclave, a pressurized steam generator used for sterilization processes.

■ Figure 15 ■

and the rickettsia that causes Q fever. Milk is pasteurized for 30 minutes at about 62°C or for 15 to 17 seconds at about 72°C. The first method is known as the **holding method,** the second method as the **flash method.** Dairy products can be pasteurized at 82°C for three seconds, a process known as **ultrapasteurization.**

An alternative heating method is **tyndallization,** also called **intermittent sterilization.** Liquids and other items are subjected to free-flowing steam for 30 minutes on each of three successive days. During the first day, all vegetating microorganisms, except spores, are killed. In the overnight period, the spores germinate, and they are killed by the steam on the second day. The last few remaining spores germinate on the second evening and are killed on the third day.

Nonheat methods. A number of **nonheat methods** are also available to control the growth and presence of microorganisms. Among these is **filtration,** a process in which a liquid or gas passes through a series of pores small enough to retain microorganisms. A vacuum can be created to help pull the liquid or gas through the filter. A filter is often used when heat-sensitive materials such as vaccines are to be sterilized.

Filter materials can be of various types. For example, certain filters consist of **diatomaceous earth,** the skeletal remains of diatoms. **Membrane filters** composed of nitrocellulose can also be used. The effectiveness of the filter depends upon the pore size, which can be established to trap the microorganisms desired. For instance, if bacteria are to be removed, the pore size would be about 0.15 μm, while if viruses are to be removed, the pores size should be about 0.01 μm.

Drying can be used to control the growth of microorganisms because when water is removed from cells, they shrivel and die. To dry foods, they are mixed with salt or sugar. Either draws water out of microbial cells by osmosis, and they quickly die. One method for achieving drying is **lyophilization,** a process in which liquids are quick-frozen and then subjected to evacuation, which dries the material. Salted meat and sugared fruits are preserved this way.

Cold temperatures are used in the refrigerator to control microbial growth. At low temperatures, microbial metabolism slows considerably, and the reproductive rate is reduced. However, cold temperatures do not necessarily kill microorganisms. At freezing temperatures, ice crystals kill many microorganisms present.

Radiations are also used to control microorganisms when food or other materials are subjected to gamma rays or X rays. The radiations change the chemical composition of microorganisms by forming ions in the organic materials of the cytoplasm. Highly reactive toxic radicals also form.

Nonionizing radiations are typified by **ultraviolet light.** Ultraviolet light affects the nucleic acids of microorganisms, inducing adjacent thymine residues in DNA molecules to bind to one another forming dimers. This binding changes the character of the DNA, making it unable to function in protein synthesis. Cell death soon follows.

Although **microwaves** are a form of radiation, their direct effect on microorganisms is minimal. Microwaves induce water molecules to vibrate at high rates, creating heat. The heat is the killing agent rather than the microwaves.

Chemical Methods of Control

Chemical agents are generally not intended to achieve sterilization. Most reduce the microbial populations to safe levels or remove pathogens from objects. An ideal disinfectant or antiseptic (chemical agent) kills microorganisms in the shortest possible time without damaging the material treated.

Among the important criteria for selecting an antiseptic or disinfectant are the concentration of disinfectant to be used, whether the agent is bactericidal or bacteriostatic, the nature of the material to be treated, whether organic matter will be present, the temperature and pH at which the chemical agent will be used, and the time available in which the chemical agent will be left in contact with the surface tested.

Evaluation methods. To evaluate an antiseptic or disinfectant, the **phenol coefficient test** is used. In this test, various dilutions of the chemical agent are prepared and tested against equivalent dilutions of phenol with such bacteria as *Staphylococcus aureus* and *Salmonella typhi*. A phenol coefficient (PC) greater than one indicates that the chemical agent is more effective than phenol and less than one that it is less effective.

An alternative test is the **in-use test.** Various dilutions of the chemical agent are made and tested against a standardized preparation of test bacteria on the type of material later to be disinfected in normal use.

Phenol. One of the first chemicals to be used for disinfection was **phenol.** First used by Joseph Lister in the 1860s, it is the standard for most other antiseptics and disinfectants. Phenol derivatives called **phenolics** contain altered molecules of phenol useful as antiseptics and disinfectants. The phenolics damage cell membranes and inactivate enzymes of microorganisms, while denaturing their proteins. They include **cresols,** such as Lysol, as well as several **bisphenols,** such as hexachlorophene, which is particularly effective against staphylococci (Figure 16a).

A chemical agent resembling the phenols is **chlorhexidine** (Hibiclens), which is used for skin disinfection as an alternative to hexachlorophene. It persists on the skin and is effective against vegetating bacteria, but not spores.

Halogens. Among the **halogen** antiseptics and disinfectants are chlorine and iodine. **Iodine** is used as a tincture of iodine, an alcohol solution. Combinations of iodine and organic molecules are called **iodophors.** They include Betadine and Isodyne, both of which contain a surface active agent called povidone. Iodine combines with microbial proteins and inhibits their function.

Chlorine also combines with microbial proteins. It is used as sodium hypochlorite (bleach). As calcium hypochlorite, chlorine is

Phenolics

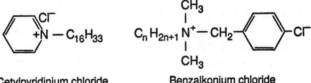

Phenol	Orthocresol	Orthophenylphenol	Hexachlorophene

(a)

Alcohols

$$CH_3-CH_2-OH \qquad CH_3-\overset{\displaystyle OH}{\overset{\displaystyle |}{CH}}-CH_3$$

Ethanol   Isopropanol

(b)

Quaternary ammonium compounds

Cetylpyridinium chloride   Benzalkonium chloride

(c)

Aldehydes

Formaldehyde   Glutaraldehyde

(d)

Gases

Ethylene oxide   Betapropiolactone

(e)

A selection of chemical disinfectants and antiseptics.

■ Figure 16 ■

available to disinfect equipment in dairies, slaughterhouses, and restaurants. The chloramines contain chlorine together with ammonia. They are used to sanitize glassware and eating utensils and are effective in the presence of organic matter. Chlorine is also used as a gas to maintain a low microbial count in drinking water.

Alcohols. Alcohols are useful chemical agents when employed against bacteria and fungi, but they have no effect on bacterial spores. The type of alcohol most widely used is 70 percent **ethyl alcohol** (ethanol). **Isopropyl alcohol** (rubbing alcohol) is also useful as an antiseptic and disinfectant. Because alcohols evaporate quickly, they leave no residue and are useful in degerming the skin before injections (Figure 16b).

Heavy metals. A number of **heavy metals** have antimicrobial ability. For example, **silver** is used as silver nitrate in the eyes of newborns to guard against infection by *Neisseria gonorrheae*. It is also used to cauterize wounds. **Copper** is used as copper sulfate to retard the growth of algae in swimming pools, fish tanks, and reservoirs. **Zinc** is useful as zinc chloride in mouthwashes and as zinc oxide as an antifungal agent in paints. The heavy metals are believed to act by combining with sulfhydryl groups on cellular proteins.

Soaps and detergents. Soaps and detergents decrease the surface tension between microorganisms and surfaces, and thereby help cleanse the surface. **Soaps** emulsify the oily film on the body surface, carrying the oils, debris, and microorganisms away in a degerming action. The cationic **detergents** are **quaternary ammonium compounds.** They solubilize the cell membranes of microorganisms. Among the popular compounds are Zephiran (benzalkonium chloride) and Cepacol (cetylpyridinium chloride) (Figure 16c).

Aldehydes. Two **aldehydes,** formaldehyde and glutaraldehyde, inactivate microbial proteins by crosslinking the functional groups in the proteins. **Formaldehyde** gas is commonly used as formalin, a 37 percent solution of formaldehyde gas. It is widely employed for embalming purposes. **Glutaraldehyde** is used as a liquid to sterilize hospital equipment. However, several hours are required to destroy bacterial spores (Figure 16d).

Ethylene oxide. Sterilization can be achieved with a chemical known as **ethylene oxide (ETO).** This chemical denatures proteins and destroys all microorganisms, including bacterial spores. It is used at warm temperatures in an ethylene oxide chamber. Several hours are needed for exposure and flushing out the gas, which can be toxic to humans. ETO is widely used for plastic instruments such as Petri dishes, syringes, and artificial heart valves (Figure 16e). **Propylene oxide,** a similar compound, is also valuable as a sterilant.

Oxidizing agents. Oxidizing agents such as **hydrogen peroxide** kill microorganisms by releasing large amounts of oxygen, which contributes to the alteration of microbial enzymes. Hydrogen peroxide is useful on inanimate objects and in foods, but on the skin surface, it is quickly broken down by the enzyme catalase, liberating oxygen. This oxygen causes a wound to bubble and thereby removes microorganisms present. However, the chemical activity on the skin is limited compared to that on inanimate surfaces. Contact lenses can be disinfected with hydrogen peroxide.

Two other oxidizing agents are **benzoyl peroxide** and **ozone.** Benzoyl peroxide is applied to the skin to treat acne due to anaerobic bacteria. The oxygen released by the compound inhibits anaerobic growth. Ozone can be used to disinfect water, where it oxidizes the cellular components of contaminating microbes.

Food preservatives. Foods can be preserved by using a number of **organic acids** to maintain a low microbial population. Sorbic acid is used in a number of acidic foods, including cheese, to prevent microbial growth. Benzoic acid also inhibits fungi and is used in acidic foods and soft drinks. Calcium propionic acid prevents the growth of mold in breads and bakery products.

Antibiotics

Various families of antibiotics are used for various types of microorganisms to achieve control and assist body defenses during times of infection. **Antibiotics** are products of microorganisms that react with and inhibit the growth of other microorganisms. An antibiotic should be selectively toxic to pathogenic microorganisms, should not incite an allergic response in the body, should not upset the normal microbial population of various body sites, and should not foster the development of drug resistance.

Penicillin. **Penicillin** prevents Gram-positive bacteria from forming peptidoglycan, the major component of the cell wall. Without peptidoglycan, internal pressures cause the bacterium to swell and burst.

Penicillin is not one antibiotic, but a family of antibiotics. The family includes penicillin F, penicillin G, and penicillin X, as well as ampicillin, amoxicillin, nafcillin, and ticarcillin. The first penicillin was derived from the green mold *Penicillium*, but most penicillins are now produced by synthetic means. A few are used against Gram-negative bacteria.

People allergic to penicillin may suffer localized allergy reactions or whole body reactions known as anaphylaxis. Long-term use of penicillin encourages the emergence of penicillin-resistant bacteria because these bacteria produce penicillinase, an enzyme that converts penicillin to penicilloic acid.

Cephalosporin antibiotics. **Cephalosporin antibiotics** include cefazolin, cefoxitin, cefotaxime, cefuroxime, and moxalactam. The antibiotics were first produced by the mold *Cephalosporium*. They prevent synthesis of bacterial cell walls, and most are useful against Gram-positive bacteria; the newer cephalosporin antibiotics are also effective against Gram-negative bacteria. Cephalosporins are especially useful against penicillin-resistant bacteria and are often used as substitutes for penicillin.

Aminoglycoside antibiotics. The **aminoglycoside antibiotics** inhibit protein synthesis in Gram-negative bacteria. Members of this antibiotic group include gentamicin, kanamycin, tobramycin, and streptomycin. Originally isolated from members of the bacterial genus *Streptomyces*, the aminoglycosides are now produced synthetically or semisynthetically. Streptomycin is effective against the tuberculosis bacterium. Unfortunately, many aminoglycosides have a deleterious effect on the ear and impair hearing.

Tetracycline antibiotics. **Tetracycline antibiotics** are broad-spectrum drugs that inhibit the growth of Gram-negative bacteria, rickettsiae, chlamydiae, and certain Gram-positive bacteria. They accomplish this by inhibiting protein synthesis. Compared to other antibiotics, tetracyclines have relatively mild side effects, but they are known to destroy helpful bacteria in the body. Also, they interfere with calcium deposit in the body, so they should not be used in very young children. Originally isolated from members of the genus *Streptomyces*, the tetracyclines include such antibiotics as minocycline, doxycycline, and tetracycline.

Other antibacterial antibiotics. The antibiotic **erythromycin** may be used as a substitute for penicillin when penicillin sensitivity or penicillin allergy exists. Erythromycin is useful against Gram-positive bacteria and has been found effective against the organisms that cause Legionnaires' disease and mycoplasmal pneumonia. It inhibits protein synthesis.

Tuberculosis is a difficult disease to treat because the etiologic agent is the extremely resistant bacterium *Mycobacterium tuberculosis*. Five drugs are currently useful for treating tuberculosis: **rifampin, ethambutol, streptomycin, para-aminosalicylic acid,** and **isoniazid.** Rifampin is also used to treat bacterial meningitis.

Bacitracin is used for the treatment of skin infections due to Gram-positive bacteria. This antibiotic inhibits cell wall synthesis in bacteria and can be used internally, but it may cause kidney damage.

Vancomycin is currently used against bacteria displaying resistance to penicillin, cephalosporin, and other antibiotics. Vancomycin is a very expensive antibiotic with numerous side effects, and it is used only in life-threatening situations. It interferes with cell wall formation in bacteria.

Chloramphenicol is effective against a broad range of bacteria including Gram-positive and Gram-negative bacteria, rickettsiae, and chlamydiae. However, it has serious side effects such as aplastic anemia (blood cells without hemoglobin), and it may induce the gray syndrome (a type of cardiovascular collapse) in babies. Therefore, it is used for only the most serious bacterial infections such as typhoid fever and meningitis.

Sulfa drugs such as **sulfamethoxazole** and **sulfisoxazole** are effective against Gram-positive bacteria. These bacteria produce folic acid, and the sulfa drugs interfere with its production by replacing para-aminobenzoic acid (PABA) in the folic acid molecule. This action typifies how an antibiotic can interfere with a metabolic pathway in bacteria.

Antifungal drugs. Several **antifungal antibiotics** are currently available for treating infectious disease. One example is **griseofulvin,** which is used against the fungi of ringworm and athlete's foot. Other examples are **nystatin, clotrimazole, ketoconazole,** and **miconazole,** all of which are used against vaginal infections due to *Candida albicans*. For systemic fungal infections, the antibiotic **amphotericin B** is available, although it has serious side effects.

Antiviral drugs. Antiviral drugs are not widely available because viruses have few functions or structures with which drugs can interfere. Nevertheless, certain drugs are available to interfere with viral replication. One example is **azidothymidine (AZT)**, which is used to interrupt the replication of human immunodeficiency virus. Other examples are **acyclovir,** which is used against herpes viruses and chickenpox viruses; **ganciclovir,** which is used against cytomegalovirus; **amantadine,** which is prescribed against influenza viruses; and **interferon,** which has been used against rabies viruses and certain cancer viruses.

Antiprotozoal drugs. Many antibiotics used against bacteria, for example, tetracycline, are also useful against protozoa. Among the drugs used widely as antiprotozoal agents are metronidazole (Flagyl), which is used against *Trichomonas vaginalis;* quinine, which is used against malaria; and pentamidine isethionate, which is valuable against *Pneumocystis carinii.*

Drug resistance. Over the past decades, **drug-resistant** strains have developed in bacteria. These strains probably existed in the microbial population, but their resistance mechanisms were not needed because the organisms were not confronted with the antibiotic. With widespread antibiotic use, the susceptible bacteria died off, and the resistant bacteria emerged. They multiplied to form populations of drug-resistant microorganisms.

Microorganisms can exhibit their resistance in various ways. For example, they can release enzymes (such as penicillinase) to inactivate the antibiotic before the antibiotic kills the microorganism; or they can stop producing the drug-sensitive structure or modify the structure so that it is no longer sensitive to the drug; or they can change the structure of the plasma membrane so that the antibiotic cannot pass to the cytoplasm.

Microorganisms have the ability to acquire genes and thereby undergo the process of **recombination.** In recombination, a new chromosome with a genotype different from that of the parent results from the combination of genetic material from two organisms. This new arrangement of genes is usually accompanied by new chemical or physical properties.

In microorganisms, several kinds of recombination are known to occur. The most common form is **general recombination,** which usually involves a reciprocal exchange of DNA between a pair of DNA sequences. It occurs anywhere on the microbial chromosome and is typified by the exchanges occurring in bacterial transformation, bacterial recombination, and bacterial transduction.

A second type of recombination, called **site-specific recombination,** involves the integration of a viral genome into the bacterial chromosome. A third type is **replicative recombination,** which is due to the movement of genetic elements as they switch position from one place on the chromosome to another.

The principles of recombination apply to prokaryotic microorganisms but not to eukaryotic microorganisms. Eukaryotes exhibit a complete sexual life cycle, including meiosis. In this process, new combinations of a particular gene form during the process of crossing over. This process occurs between homologous chromosomes and is not seen in bacteria, where only a single chromosome exists. Much of the work in microbial genetics has been performed with bacteria, and the unique features of microbial genetics are usually those associated with prokaryotes such as bacteria.

The Bacterial Chromosome and Plasmid

While eukaryotes have two or more chromosomes, prokaryotes such as bacteria possess a single **chromosome** composed of double-stranded DNA in a loop. The DNA is located in the nucleoid of the cell and is not associated with protein. In *Escherichia coli,* the length of the chromosome, when open, is many times the length of the cell.

Many bacteria (and some yeasts or other fungi) also possess looped bits of DNA known as **plasmids,** which exist and replicate independently of the chromosome. Plasmids have relatively few genes (fewer than 30). The genetic information of the plasmid is usually not essential to survival of the host bacteria.

Plasmids can be removed from the host cell in the process of **curing.** Curing may occur spontaneously or may be induced by treatments such as ultraviolet light. Certain plasmids, called **episomes,** may be integrated into the bacterial chromosome. Others contain genes for certain types of pili and are able to transfer copies of themselves to other bacteria. Such plasmids are referred to as **conjugative plasmids.**

A special plasmid called a **fertility (F) factor** plays an important role in conjugation. The F factor contains genes that encourage cellular attachment during conjugation and accelerate plasmid transfer between conjugating bacterial cells. Those cells contributing DNA are called F^+ **(donor) cells,** while those receiving DNA are the F^- **(recipient) cells.** The F factor can exist outside the bacterial chromosome or may be integrated into the chromosome.

Plasmids contain genes that impart antibiotic resistance. Up to eight genes for resisting eight different antibiotics may be found on a single plasmid. Genes that encode a series of **bacteriocins** are also found on plasmids. Bacteriocins are bacterial proteins capable of destroying other bacteria. Still other plasmids increase the pathogenicity of their host bacteria because the plasmid contains genes for toxin synthesis.

Transposable elements. Transposable elements, also known as **transposons,** are segments of DNA that move about within the chromosome and establish new genetic sequences. First discovered by Barbara McClintock in the 1940s, transposons behave somewhat like lysogenic viruses except that they cannot exist apart from the chromosome or reproduce themselves.

The simplest transposons, **insertion sequences,** are short sequences of DNA bounded at both ends by identical sequences of nucleotides in reverse orientation (inverted repeats). Insertion sequences can insert within a gene and cause a rearrangement mutation of the genetic material. If the sequence carries a stop codon, it may block transcription of the DNA during protein synthesis. Insertion sequences may also encourage the movement of drug-resistance genes between plasmids and chromosomes.

Bacterial Recombinations

Three types of bacterial recombination result in a change in the DNA of recipient organisms. The proteins expressed by the new genes lead to new physiological characteristics in the bacteria.

Bacterial conjugation. Bacterial **conjugation** was first postulated in the 1940s by Joshua Lederberg and Edward Tatum. The essential feature of the process is that two bacterial cells come together and mate such that a gene transfer occurs between them. One cell, the donor cell (F^+), gives up DNA; and another cell, the recipient cell (F^-), receives the DNA. The transfer is nonreciprocal, and a special pilus called the **sex pilus** joins the donor and recipient during the transfer. The DNA most often transferred is a copy of the F factor plasmid. The factor moves to the recipient, and when it enters the recipient, it is copied to produce a double-stranded DNA for integration. The channel for transfer is usually a special **conjugation tube** formed during contact between the two cells (Figure 17).

Certain donor strains of bacteria transfer genes with high efficiency. In this case, the F factor acts as an episome and integrates itself into the bacterial chromosome. Under these conditions, chromosomal genes are transferred to the recipient cell, and the donor is called a **high frequency of recombination (Hfr) donor.** During normal conjugation, the donor cell can become a recipient cell if the F factor is transferred during the conjugation. However, during Hfr conjugation, the F factor is rarely transmitted, and the recipient cell

F-factor Transfer

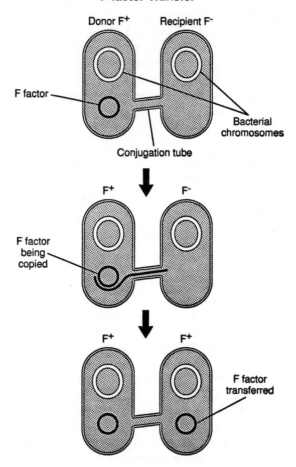

The process of bacterial conjugation using the F factor plasmid.

■ Figure 17 ■

does not become a donor cell. The exception occurs if the complete chromosome is transferred, a process requiring about 100 minutes in *E. coli*. In this case, the F factor is transferred and the recipient becomes a donor cell.

During some instances of conjugation, the F plasmid leaves the bacterial chromosome carrying an excised piece of chromosomal DNA The plasmid carrying the chromosomal DNA is called an **F′ plasmid.** If the F′ plasmid is transferred to a recipient gene during conjugation, the donor bacterial genes will also be transferred. This type of conjugation is important because it accounts for the spread of certain bacterial genes through a bacterial population. The process is called **sexduction.**

Bacterial transformation. Bacterial **transformation** was discovered by Frederick Griffith in 1928. Griffith worked with the **pneumococci** that cause bacterial pneumonia. He discovered that if he mixed fragments of dead pathogenic pneumococci with specimens of live harmless pneumococci, the harmless bacteria took on genes of the bacterial fragments and became pathogenic. Griffith's work with pneumococci was among the first demonstrating that bacteria could undergo genetic changes.

Scientists now recognize that when bacteria undergo lysis, they release considerable amounts of DNA into the environment. This DNA may be picked up by a **competent cell,** that is, one capable of taking up the DNA and undergoing a transformation. To be competent, bacteria must be in the logarithmic stage of growth, and a **competence factor** needed for the transformation must be present.

During transformation, a competent cell takes up DNA and destroys one strand of the double helix. A single-stranded fragment then replaces a similar but not identical fragment in the recipient organism, and the transformation is complete. Transformation has been studied in detail in *Streptococcus pneumoniae* and *Haemophilus influenzae*. It can be encouraged in the laboratory by treating cells with heat and calcium chloride, a process that increases the permeability of the cell membrane to DNA.

Bacterial transduction. The third important kind of bacterial recombination is **transduction.** In transduction, **bacterial viruses** (also known as **bacteriophages**) transfer DNA fragments from one bacterium (the donor) to another bacterium (the recipient). The viruses involved contain a strand of DNA enclosed in an outer coat of protein.

After a bacteriophage (or phage, in brief) enters a bacterium, it may encourage the bacterium to make copies of the phage. At the conclusion of the process, the host bacterium undergoes lysis and releases new phages. This cycle is called the **lytic cycle.** Under other circumstances, the virus may attach to the bacterial chromosome and integrate its DNA into the bacterial DNA. It may remain here for a period of time before detaching and continuing its replicative process. This cycle is known as the **lysogenic cycle.** Under these conditions, the virus does not destroy the host bacterium, but remains in a lysogenic condition with it. The virus is called a **temperate phage,** also known as a **prophage.** At a later time, the virus can detach, and the lytic cycle will ensue.

During **generalized transduction,** a phage assumes a lysogenic condition with a bacterium, and the phage DNA remains with the chromosomal DNA. When the phage replicates, however, random fragments of the bacterial DNA are packaged in error by new phages during their production. The result is numerous phages containing genes from the bacterium in addition to their own genes. When these phages enter a new host bacterium and incorporate their DNA to the bacterial chromosome, then they will also incorporate the DNA from the previous bacterium, and the recipient bacterium will be transduced (Figure 18). It will express not only its genes, but also the genes acquired from the donor bacterium.

A second type of transduction is called **specialized transduction.** In this case, the lysogenic cycle ensues as before. When the phage DNA breaks away from the bacterial DNA, however, it may take with it a small amount of the bacterial DNA (perhaps 5 percent). When the phage DNA is used as a template for the synthesis of new phage DNA particles, the bacterial genes are also reproduced. When the phages enter new bacterial cells, they carry the bacterial genes along with them. In the recipient bacterium, the phage and donor genes integrate into the bacterial chromosome and transduce the recipient organism. Specialized transduction is an extremely rare event in comparison to generalized transduction because genes do not easily break free from the bacterial chromosome.

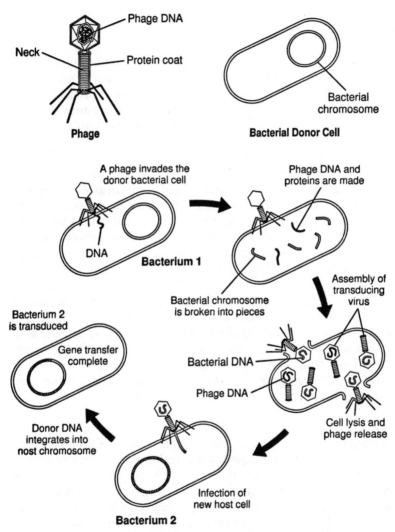

Generalized transduction involving a bacterial virus (bacteriophage) and a donor bacterium.

■ Figure 18 ■

Mutation

A **mutation** is a permanent alteration in the sequence of nitrogenous bases of a DNA molecule. The result of a mutation is generally a change in the end-product specified by that gene. In some cases, a mutation can be beneficial if a new metabolic activity arises in a microorganism, or it can be detrimental if a metabolic activity is lost.

Types of mutations. The most common type of mutation involves a single base pair in the DNA molecule and is known as a **point mutation.** In this case, a different base is substituted for the normal base, thus altering the genetic code. Should a new amino acid be substituted in the final protein, the mutation is known as **missense mutation.** Certain mutations change the genetic code and destroy the information it contains. Such a mutation is referred to as a **nonsense mutation.**

In another type of cell mutation, a **frameshift mutation,** pairs of nucleotides are either added to or deleted from the DNA molecule, with the result that the "reading frame" is shifted. The amino acid sequence in the resulting protein changes as a result of this frameshift. If a mutation occurs without laboratory intervention, it is a **spontaneous mutation;** if it occurs as a result of laboratory intervention, it is an **induced mutation.**

Mutagens. Physical and chemical agents capable of bringing about mutations are called **mutagens. Chemical mutagens** include nitrous acid. This substance converts adenine to hypoxanthine, a molecule that will not pair with thymine, and thus interrupts the genetic code. A **base analog** is a chemical mutagen that resembles a nitrogenous base and is incorporated by error into a DNA molecule. Such a DNA molecule cannot function in protein synthesis. Certain dyes and fungal toxins (for example, aflatoxin) are known to be mutagens.

Physical mutagens include X rays, gamma rays, and ultraviolet light. X rays and gamma rays break the covalent bonds in DNA molecules, thereby producing fragments. Ultraviolet light binds together adjacent thymine bases, forming dimers. These dimers cannot function in protein synthesis, and the genetic code is thereby interrupted. Radiation damage can be repaired by certain bacterial enzymes, a process known as **photoreactivation.**

The probability of a mutation occurring during cellular division is known as the **mutation rate.** In bacteria, the spontaneous mutation rate is about one in a billion reproductions. This factor implies that in every population of a billion cells, there is at least one mutant. This mutant organism may never express its mutation. However, for example, if the mutation renders antibiotic resistance, then the mutants will survive when an antibiotic is applied to the population, and a new colony of antibiotic-resistant bacteria will emerge.

During the 1950s, a tremendous explosion in biological research occurred, and the methods of gene expression were elucidated. The knowledge generated during this period helped explain how genes function in microorganisms and gave rise to the science of molecular genetics. This science is concerned with the activity of deoxyribonucleic acid (DNA) and how that activity brings about the production of proteins in microbial and other cells.

DNA Structure

As proposed originally in 1953 by Watson and Crick, **deoxyribonucleic acid (DNA)** consists of two long chains of nucleotides. The two nucleotide chains twist around one another to form a **double helix,** which resembles a spiral staircase. The two chains of nucleotides are held to one another by weak hydrogen bonds between bases of the chains.

A **nucleotide** in the DNA chain consists of three parts: a nitrogenous base, a phosphate group, and a molecule of deoxyribose. The **nitrogenous bases** of each nucleotide chain are of two major types: purines and pyrimidines. **Purines** have two fused rings of carbon and nitrogen atoms, while **pyrimidines** have only one ring. The two purine bases in DNA are **adenine (A)** and **guanine (G)**. The pyrimidine bases in DNA are **cytosine (C)** and **thymine (T)**. Purine and pyrimidine bases are found in both strands of the double helix.

The **phosphate group** of DNA is derived from a molecule of phosphoric acid and connects the deoxyribose molecules to one another in the nucleotide chain. **Deoxyribose** is a five-carbon carbohydrate. The purine and pyrimidine bases are attached to the deoxyribose molecules and stand opposite one another on the two nucleotide chains. Adenine always stands opposite and binds to thymine. Guanine always stands opposite and binds to cytosine. Adenine and thymine are said to be complementary, as are guanine and cytosine. This is known as the principle of **complementary base pairing.**

DNA replication. Before a cell enters the process of binary fission or mitosis, the DNA replicates itself to ensure that the daughter cells can function independently. In the process of **DNA replication,** specialized enzymes pull apart, or "unzip," the DNA double helix.

As the two strands separate, the purine and pyrimidine bases on each strand are exposed. The exposed bases then attract their complementary bases and induce the complementary bases to stand opposite. Deoxyribose molecules and phosphate groups are brought into the environment, and the enzyme DNA polymerase unites all the nucleotide components to one another and forms a long strand of nucleotides. Thus, the old strand of DNA directs the synthesis of a new strand of DNA through complementary base pairing.

After the synthesis has occurred, one old strand of DNA unites with a new strand to reform a double helix. This process is called **semiconservative replication** because one of the old strands is conserved in the new DNA double helix.

Protein Synthesis

During the 1950s and 1960s it became apparent that DNA is essential in the synthesis of proteins. Proteins are used as structural materials in the cells and function as enzymes. In addition, many specialized proteins function in cellular activities. For example, in bacteria, flagella and pili are composed of protein.

The genetic code. The key element of a protein molecule is how the amino acids are linked. The sequences of amino acids, determined by genetic codes in DNA, distinguish one protein from another. The **genetic code** consists of the sequence of nitrogenous bases in the DNA. How the nitrogenous base code is translated to an amino acid sequence in a protein is the basis for protein synthesis.

In order for protein synthesis to occur, several essential materials must be present. One is a supply of the 20 amino acids which make up most proteins. Another essential element is a series of enzymes that will function in the process. DNA and another form of nucleic acid called **ribonucleic acid (RNA)** are also essential. RNA carries instructions from the nuclear DNA into the cytoplasm, where protein is synthesized. RNA is similar to DNA, with three exceptions. First, the carbohydrate in RNA is ribose rather than deoxyribose. Second, RNA nucleotides contain the pyrimidine **uracil** rather than thymine. And third, RNA is usually single-stranded.

Types of RNA. In the synthesis of protein, three types of RNA are required. The first is called **ribosomal RNA (rRNA)** and is used to manufacture ribosomes. **Ribosomes** are ultramicroscopic particles of rRNA and protein where amino acids are linked to one another during the synthesis of proteins. Ribosomes may exist along the membranes of the endoplasmic reticulum in eukaryotic cells or free in the cytoplasm of prokaryotic cells .

A second important type of RNA is **transfer RNA (tRNA)**, which is used to carry amino acids to the ribosomes for protein synthesis. Molecules of tRNA exist free in the cytoplasm of cells. When protein synthesis is taking place, enzymes link tRNA to amino acids in a highly specific manner.

The third form of RNA is **messenger RNA (mRNA)**, which receives the genetic code from DNA and carries it into the cytoplasm where protein synthesis takes place. In this way, a genetic code in the DNA can be used to synthesize a protein at a distant location at the ribosome. The synthesis of mRNA, tRNA, and rRNA is accomplished by an enzyme called **RNA polymerase.**

Transcription. **Transcription** is one of the first processes in the overall process of protein synthesis. In transcription, a strand of mRNA is synthesized using the genetic code of DNA. RNA polymerase binds to an area of a DNA molecule in the double helix (the other strand remains unused). The enzyme moves along the DNA strand and selects complementary bases from available nucleotides and positions them in an mRNA molecule according to the principle of complementary base pairing (Figure 19). The chain of mRNA lengthens until a stop code is received.

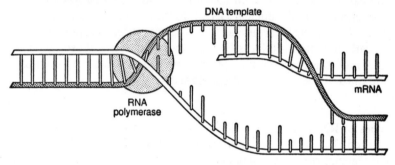

The synthesis of mRNA using a strand of DNA as a template.

■ Figure 19 ■

The nucleotides of the DNA strands are read in groups of three. Each triplet is called a **codon.** Thus, a codon may be CGA, or TTA, or GCT, or any other combination of the four bases, depending on their sequence in the DNA strand. The mRNA molecule consists of a series of codons received from the genetic message in the DNA.

Once the stop codon has been reached, the mRNA molecule leaves the DNA molecule, and the DNA molecule rewinds to form a double helix. Meanwhile, the mRNA molecule proceeds thorough the cellular cytoplasm toward the ribosomes.

Translation. **Translation** is the process in which the genetic code will be "translated" to an amino acid sequence in a protein. The process begins with the arrival of the mRNA molecule at the ribosomes. While mRNA was being synthesized, tRNA molecules were uniting with their specific amino acids according to the activity of specific enzymes. The tRNA molecules then began transporting their amino acids to the ribosomes to meet the mRNA molecule.

After it arrives at the ribosomes, the mRNA molecule exposes its bases in sets of three, the codons. Each codon has a complementary codon called an **anticodon** on a tRNA molecule. When the codon of the mRNA molecule complements the anticodon on a tRNA molecule, the latter places the particular amino acid in that position. Then the next codon of the mRNA is exposed, and the complementary anticodon of a tRNA molecule matches with it. The amino acid carried by the second tRNA molecule is thus positioned next to the first amino acid, and the two are linked. At this point, the tRNA molecules release their amino acids and return to the cytoplasm to link up with new molecules of amino acid.

The ribosome then moves farther down the mRNA molecule and exposes another codon which attracts another tRNA molecule with its anticodon. Another amino acid is brought into position. In this way, amino acids continue to be added to the growing chain until the ribosome has moved down to the end of the mRNA molecule. The sequence of codons on the mRNA molecule thus determines the sequence of amino acids in the protein being constructed (Figure 20).

Once the protein has been completely synthesized, it is removed from the ribosome for further processing. For example, the protein may be stored in the Golgi body of a eukaryotic cell before release, or a bacterium may release it as a toxin. The mRNA molecule is broken up and the nucleotides are returned to the nucleus. The tRNA molecules return to the cytoplasm to unite with fresh molecules of amino acids, and the ribosome awaits the arrival of a new mRNA molecule.

Gene control. The process of protein synthesis does not occur constantly in the cell, but rather at intervals followed by periods of genetic "silence." Thus, the process of gene expression is regulated and controlled by the cell.

The control of gene expression can occur at several levels in the cell. For example, genes rarely operate during mitosis. Other levels of gene control can occur at transcription, when certain segments of DNA increase and accelerate the activity of nearby genes. After transcription has taken place, the mRNA molecule can be altered to regulate gene activity. For example, it has been found that eukaryotic

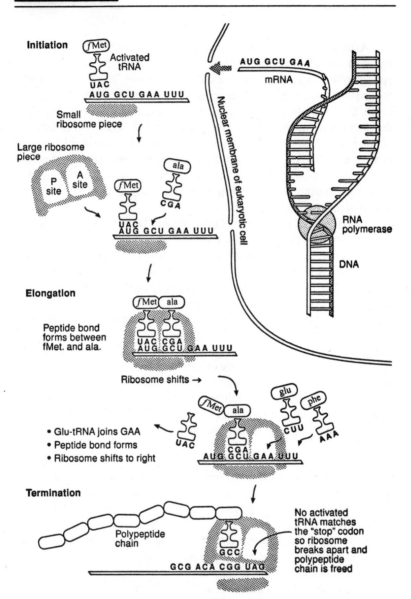

Steps in the synthesis of protein beginning with the genetic code in DNA and ending with the finished polypeptide chain.

■ Figure 20 ■

mRNA contains many useless bits of RNA that are removed in the production of the final mRNA molecule. These useless bits of nucleic acid are called **introns.** The remaining pieces of mRNA, called **exons,** are then spliced to form the final mRNA molecule. Bacterial mRNA lacks introns.

The concept of gene control has been researched thoroughly in bacteria. In these microorganisms, genes have been identified as structural genes, regulator genes, and control regions. The three units form a functional unit called the **operon.**

The operon has been examined in close detail in certain bacteria. It has been found that certain carbohydrates can induce the presence of the enzymes needed to digest those carbohydrates. For example, when lactose is present, bacteria synthesize the enzymes needed to break it down. Lactose acts as an **inducer molecule** in the following way: In the absence of lactose, a regulator gene produces a repressor protein, which binds to a control region called the **operator site.** This binding prevents the structural genes from encoding the enzyme for lactose digestion. When lactose is present, however, it binds to the repressor protein and thereby removes the repressor at the operator site. With the operator site free, the structural genes are released to produce their lactose-digesting enzyme.

Recombinant DNA and Biotechnology

Biotechnology is an industrial process that uses the scientific research on DNA for practical benefits. Biotechnology is synonymous with **genetic engineering** because the genes of an organism are changed during the process and the DNA of the organism is recombined. Recombinant DNA and biotechnology can be used to form proteins not normally produced in a cell. In addition, bacteria that carry recombinant DNA can be released into the environment to increase the fertility of the soil, serve as an insecticide, or relieve pollution.

Tools of biotechnology. The basic process of recombinant DNA technology revolves around the activity of DNA in the synthesis of protein. By intervening in this process, scientists can change the nature
of the DNA and of the gene make-up of an organism. By inserting
genes into the genome of an organism, the scientist can induce the
organism to produce a protein it does not normally produce.

The technology of recombinant DNA has been made possible in
part by extensive research on microorganisms during the last century. One important microorganism in recombinant DNA research is
Escherichia coli (E. coli). The biochemistry and genetics of *E. coli*
are well known, and its DNA has been isolated and made to accept
new genes. The DNA can then be forced into fresh cells of *E. coli,*
and the bacteria will begin to produce the proteins specified by the
foreign genes. Such altered bacteria are said to have been transformed.

Interest in recombinant DNA and biotechnology heightened considerably in the 1960s and 1970s with the discovery of **restriction
enzymes.** These enzymes catalyze the opening of a DNA molecule at
a "restricted" point, regardless of the DNA's source. Moreover, certain restriction enzymes leave dangling ends of DNA molecules at
the point where the DNA is open. (The most commonly used restriction enzyme is named *Eco*RI.) Foreign DNA can then be combined
with the carrier DNA at this point. An enzyme called **DNA ligase** is
used to form a permanent link between the dangling ends of the DNA
molecules at the point of union (Figure 21).

The genes used in DNA technology are commonly obtained from
host cells or organisms called **gene libraries.** A gene library is a collection of cells identified as harboring a specific gene. For example,
E. coli cells can be stored with the genes for human insulin in their
chromosomes.

Pharmaceutical products. Gene defects in humans can lead to deficiencies in proteins such as insulin, human growth hormone, and Factor VIII. These protein deficiencies may lead to problems such as
diabetes, dwarfism, and impaired blood clotting, respectively. Miss-
ing proteins can now be replaced by proteins manufactured through
biotechnology. For **insulin** production, two protein chains are encoded

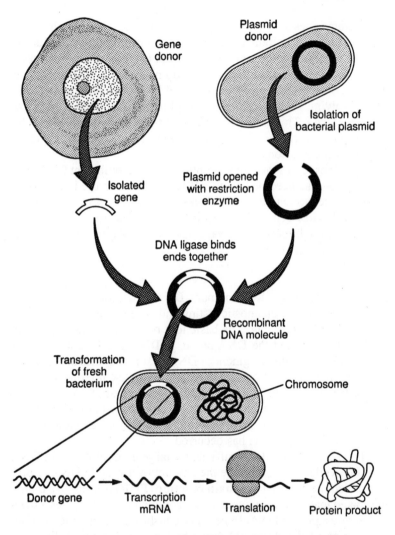

The production of a a recombined bacterium using a gene from a foreign donor and the synthesis of protein encoded by the recombinant DNA molecule.

■ Figure 21 ■

by separate genes in plasmids inserted into bacteria. The protein chains are then chemically joined to form the final insulin product. **Human growth hormone** is also produced within bacteria, but special techniques are used because the bacteria do not usually produce human proteins. Therapeutic proteins produced by biotechnology include a clot-dissolving protein called **tissue plasminogen activator (TPA)** and **interferon.** This antiviral protein is produced within *E. coli* cells. Interferon is currently used against certain types of cancers and for certain skin conditions.

Vaccines represent another application of recombinant DNA technology. For instance, the hepatitis B vaccine now in use is composed of viral protein manufactured by yeast cells, which have been recombined with viral genes. The vaccine is safe because it contains no viral particles. Experimental vaccines against AIDS are being produced in the same way.

Diagnostic testing. Recombinant DNA and biotechnology have opened a new era of diagnostic testing and have made detecting many genetic diseases possible. The basic tool of DNA analyses is a fragment of DNA called the DNA probe. A **DNA probe** is a relatively small, single-stranded fragment of DNA that recognizes and binds to a complementary section of DNA in a complex mixture of DNA molecules. The probe mingles with the mixture of DNA and unites with the target DNA much like a left hand unites with the right. Once the probe unites with its target, it emits a signal such as radioactivity to indicate that a reaction has occurred.

To work effectively, a sufficiently large amount of target DNA must be available. To increase the amount of available DNA, a process called the **polymerase chain reaction (PCR)** is used. In a highly automated machine, the target DNA is combined with enzymes, nucleotides, and a primer DNA. In geometric fashion, the enzymes synthesize copies of the target DNA, so that in a few hours billions of molecules of DNA exist where only a few were before.

Using DNA probes and PCR, scientists are now able to detect the DNA associated with HIV (and AIDS), Lyme disease, and genetic diseases such as cystic fibrosis, muscular dystrophy, Huntington's disease, and fragile X syndrome.

Gene therapy. **Gene therapy** is a recombinant DNA process in which cells are taken from the patient, altered by adding genes, and replaced in the patient, where the genes provide the genetic codes for proteins the patient is lacking.

In the early 1990s, gene therapy was used to correct a deficiency of the enzyme **adenosine deaminase (ADA)**. Blood cells called lymphocytes were removed from the bone marrow of two children; then genes for ADA production were inserted into the cells using viruses as vectors. Finally, the cells were reinfused to the bodies of the two children. Once established in the bodies, the gene-altered cells began synthesizing the enzyme ADA and alleviated the deficiency.

Gene therapy has also been performed with patients with **melanoma** (a virulent skin cancer). In this case, lymphocytes that normally attack tumors are isolated in the patients and treated with genes for an anticancer protein called **tumor necrosis factor.** The gene-altered lymphocytes are then reinfused to the patients, where they produce the new protein which helps destroy cancer cells. Approximately 2000 single-gene defects are believed to exist, and patients with these defects may be candidates for gene therapy.

DNA fingerprinting. The use of DNA probes and the development of retrieval techniques have made it possible to match DNA molecules to one another for identification purposes. This process has been used in a forensic procedure called **DNA fingerprinting.**

The use of DNA fingerprinting depends upon the presence of repeating base sequences that exist in the human genome. The repeating sequences are called **restriction fragment length polymorphisms (RFLPs).** As the pattern of RFLPs is unique for every individual, it can be used as a molecular fingerprint. To perform DNA fingerprinting, DNA is obtained from an individual's blood cells, hair fibers, skin fragments, or other tissue. The DNA is extracted from the cells and digested with enzymes. The resulting fragments are separated by a process called electrophoresis. These separated DNA fragments are tested for characteristic RFLPs using DNA probes. A statistical evaluation enables the forensic pathologist to compare a suspect's DNA with the DNA recovered at a crime scene and to assert with a degree of certainty (usually 99 percent) that the suspect was at the crime scene.

DNA and agriculture. Although plants are more difficult to work with than bacteria, gene insertions can be made into single plant cells, and the cells can then be cultivated to form a mature plant. The major method for inserting genes is through the plasmids of a bacterium called *Agrobacterium tumefaciens*. This bacterium invades plant cells, and its plasmids insert into plant chromosomes carrying the genes for tumor induction. Scientists remove the tumor-inducing genes and obtain a plasmid that unites with the plant cell without causing any harm.

Recombinant DNA and biotechnology have been used to increase the efficiency of plant growth by increasing the efficiency of the plant's ability to fix nitrogen. Scientists have obtained the genes for nitrogen fixation from bacteria and have incorporated those genes into plant cells. By obtaining nitrogen directly from the atmosphere, the plants can synthesize their own proteins without intervention of bacteria as normally needed.

DNA technology has also been used to increase plant resistance to disease. The genes for an insecticide have been obtained from the bacterium *Bacillus thuringiensis* and inserted into plants to allow them to resist caterpillars and other pests. In addition, plants have been reengineered to produce the capsid protein that encloses viruses. These proteins lend resistance to the plants against viral disease.

The human genome. One of the most ambitious scientific endeavors of this century is the effort to sequence the nitrogenous bases in the **human genome.** Begun in 1990, the effort is expected to encompass 15 years of work at a cost of approximately $3 billion. Knowing the content of the human genome will help researchers devise new diagnostics and treatments for genetic diseases and will also be of value to developmental biologists, evolutionary biologists, and comparative biologists.

In addition to seeking to learn the genome of humans, the project has also studied numerous bacteria. By 1995, the genomes of two bacteria had been completely deciphered (*Haemophilus influenzae* and *Mycoplasma genitalium*), and by 1996, the genome of the yeast *Saccharomyces cerevisiae* was known. The Human Genome Project is one of colossal magnitude that will have an impact on many branches of science for decades to come. The project remains the crowning achievement of DNA research in the twentieth century and the bedrock for research in the twenty-first.

Over 400 recognized genera of **bacteria** are known to exist. Bacterial species are listed in *Bergey's Manual of Systematic Bacteriology.* The entire kingdom of bacteria, including cyanobacteria, is entitled Prokaryotae. Four divisions of bacteria based on their cell wall characteristics are included in the Prokaryotae kingdom. Not all bacteria are assigned to a division, but all are assigned to one of 33 "sections."

Spirochetes and Spirilla

Spirochetes have a spiral shape, a flexible cell wall, and motility mechanisms based on structures called **axial filaments.** Each axial filament is composed of fibrils extending toward each other between two layers of the cell wall.

Spirochetes are very slender and difficult to see under the light microscope. They are cultivated with great difficulty (some cannot be cultivated), and their classification is based on their morphology and pathogenicity. Certain species inhabit water environments, while others are parasites of arthropods (such as ticks and lice) as well as warm-blooded animals. Spirochetes include *Borrelia burgdorferi*, the agent of Lyme disease, *Treponema pallidum*, the cause of syphilis, and *Leptospira interrogans*, the agent of leptospirosis.

Spirilla have a spiral shape, a rigid cell wall, and motility mechanisms based on polar flagella. The genera *Spirillum, Aquaspirillum,* and *Azospirillum* are widely dispersed among and readily isolated from numerous environments. These organisms are aerobic bacteria wound like helices. Species *S. minor* is a cause of rat bite fever in humans. The genus *Campylobacter* contains several pathogenic species, including *C. jejuni*, which causes campylobacteriosis, an intestinal infection accompanied by diarrhea.

Gram-Negative Rods and Cocci

Bdellovibrios. **Bdellovibrios** are aerobic Gram-negative, curved rods that prey on other bacteria. The organism attaches to the surface of a bacterium, rotates, and bores a hole through the host cell wall. It then takes biochemical control of the host cell and grows in the space between the cell wall and plasma membrane. The host bacterium is killed in the process. The comma-shaped *Bdellovibrio bacteriovorus* is the most thoroughly studied species of the group.

Pseudomonads. **Pseudomonads** are aerobic, Gram-negative rods that are motile with polar flagella. Over 30 species are found in the group, and *Pseudomonas fluorescens* is a well-known producer of a yellow-green pigment. Another species, *P. aeruginosa,* causes urinary tract infections and infections of burned tissue.

Azotobacter **and** ***Rhizobium.*** Species of *Azotobacter* and *Rhizobium* are extremely important for their ability to fix nitrogen in the environment. These Gram-negative rods live free in the soil *(Azotobacter)* or on the roots of legume plants *(Rhizobium)* and use their enzymes to convert atmospheric nitrogen into organic molecules useful to the plant. The plants then use the nitrogen compounds for the synthesis of amino acids and proteins, which serve as an extremely valuable food source for animals and humans. Members of the genus *Azotobacter* form a resting cell called a cyst, which withstands drying and environmental stresses.

Enterobacteria. **Enterobacteria** are facultatively anaerobic, Gram-negative rods that inhabit the human intestine. Members of the enterobacteria group are members of the family **Enterobacteriacae** classified in section 5 of *Bergey's Manual.*

Over 25 genera of enterobacteria are recognized, many with pathogenic importance. Among the medically important enterobacteria are

Salmonella species that cause intestinal disease known as salmonellosis; *Yersinia pestis,* the cause of plague; *Klebsiella* species, the causes of pneumonia, intestinal disease, and other infections; and species of *Serratia* and *Proteus.* The well-known organism *Escherichia coli* is also a member of this group. All enterobacteria have peritrichous flagella.

Vibrios. Vibrios are curved, Gram-negative, facultatively anaerobic rods. They belong to the family Vibrionaceae. One species, *Vibrio cholerae,* is the cause of cholera in humans. Members of the genus *Aeromonas* and *Plesiomonas* are involved in human intestinal disease. Species of *Photobacterium* are marine organisms known for their ability to produce light as a result of chemical actions stimulated by the enzyme luciferase. This production of light is known as bioluminescence.

Pasteurellas. The **pasteurellas** belong to the family Pasteurellaceae. They are distinguished from vibrios and enterobacteria by their small size and inability to move. The genera *Pasteurella, Haemophilus,* and *Actinobacillus* are among the important members of the group. The species *H. influenzae* is a cause of meningitis in children, while *P. multocida* causes cholera in fowl.

Sulfur bacteria. The **sulfur bacteria** use sulfur or sulfur compounds as electron acceptors in their metabolism. These bacteria produce large amounts of hydrogen sulfide during their growth, and therefore, they produce foul odors in water and mud. Members of the genus *Desulfovibrio* are particularly important in the sulfur cycle for their ability to use sulfur and convert it to other compounds that can be used by plants to synthesize sulfur-containing amino acids.

Bacteroides. The **bacteroides** are genera of anaerobic bacteria having unique motility and flagellation patterns. Several species digest cellulose in the rumen of the cow and thereby break down plants. Human feces contains large numbers of bacteria belonging to the genus *Bacteroides*, which may be helpful in digestive processes. One species, *B. fragilis,* is a possible cause of human blood infections.

Veillonella. Among the Gram-negative cocci are a group of anaerobic diplococci belonging to the genus ***Veillonella.*** *Veillonella* species are part of the normal flora of the mouth and gastrointestinal tract, and they are found in dental plaque. They are anaerobic organisms that may also cause infections of the female genital tract.

Gliding bacteria. Certain bacterial species are able to move by **gliding** in a layer of **slime,** which they produce. Wavelike contractions of the outer membranes help the bacteria propel themselves. Members of the group include species of *Cytophaga* and *Simonsiella*.

Two important genera of gliding bacteria are *Beggiatoa* and *Thiothrix*. Species of these organisms live in sulfur environments and break down hydrogen sulfide to release sulfur in the form of sulfur granules. For this reason, the bacteria are very important in the recycling of sulfur in water and soil. The bacteria are Gram-negative.

Myxobacteria are gliding bacteria that are Gram-negative, aerobic rods. They are nonphotosynthetic species and have a unique developmental cycle that involves the formation of fruiting bodies. When nutrients are exhausted, the bacteria congregate and produce a stalk, at the top of which is a mass of cells. These cells differentiate into spherical cells, similar to cysts, which are resistant to environmental extremes.

Sheathed bacteria. Sheathed bacteria are filamentous bacteria with cell walls enclosed in a **sheath** of polysaccharides and lipoproteins. The sheath assists attachment mechanisms and imparts protection to the bacteria. The genus *Sphaerotilus* is in this group.

Photoautotrophic bacteria. Photoautotrophic bacteria are Gram-negative rods which obtain their energy from sunlight through the processes of photosynthesis. In this process, sunlight energy is used in the synthesis of carbohydrates. Certain photoautotrophs called **anoxygenic photoautotrophs** grow only under anaerobic conditions and neither use water as a source of hydrogen nor produce oxygen from photosynthesis. Other photoautotrophic bacteria are **oxygenic photoautotrophs.** These bacteria are **cyanobacteria.** They use chlorophyll pigments and photosynthesis in photosynthetic processes resembling those in algae and complex plants. During the process, they use water as a source of hydrogen and produce oxygen as a product of photosynthesis.

Cyanobacteria include various types of bacterial rods and cocci, as well as certain filamentous forms. The cells contain thylakoids, which are cytoplasmic, platelike membranes containing chlorophyll. The organisms produce **heterocysts,** which are specialized cells believed to function in the fixation of nitrogen compounds.

Chemoautotrophic bacteria. Chemoautotrophic (or chemolithotrophic) bacteria are a group of Gram-negative bacteria deriving their energy from chemical reactions involving inorganic material. Certain chemoautotrophic bacteria use carbon dioxide as a carbon source and grow in a medium containing inorganic substances. By comparison, members of the genus *Thiobacillus* metabolize sulfur compounds, and members of the genera *Nitrosomonas* and *Nitrobacter* metabolize nitrogen compounds. Certain chemoautotrophic bacteria use hydrogen gas in their chemical reactions, and others use metals such as iron and manganese in their energy metabolism. These unusual types of biochemistry are characteristic of organisms found outside the body in the soil and water environment.

Gram-Positive Bacteria

Streptococci. **Streptococci** are spherical bacteria that divide in parallel planes to produce chains. The bacteria are Gram-positive, and certain species are aerobic, while others are anaerobic. On blood agar, certain species partly destroy the red blood cells and are said to be **alpha-hemolytic.** Other species completely destroy the blood cells and are **beta-hemolytic.** Those streptococci producing no blood cell destruction are **gamma-hemolytic.**

One species of streptococcus (*Streptococcus pneumoniae*) is the cause of secondary bacterial pneumonias, while another species (*Streptococcus pyogenes*) causes strep throat and rheumatic fever. Other species are associated with dental caries. Harmless strains of streptococci are used in the production of yogurt, buttermilk, and cheese.

Staphylococci. **Staphylococci** are Gram-positive bacteria that divide in planes to produce clusters or packets. Normally associated with the skin and mucous membranes, certain species of staphylococci are involved in skin boils, abscesses, and carbuncles, especially if they produce the enzyme coagulase, which causes blood clotting. *Staphylococcus aureus* is involved in cases of food poisoning, toxic shock syndrome, pneumonia, and staphylococcal meningitis.

Lactobacilli. **Lactobacilli** are Gram-positive, rod-shaped bacteria occurring as single cells or chains. They produce lactic acid in their metabolism and are associated with the flora of the mouth and the vagina. Certain species are associated with the production of dairy products such as yogurt, sour cream, and buttermilk.

***Bacillus* and *Clostridium* species.** Species of *Bacillus* and *Clostridium* are Gram-positive, rod-shaped bacteria able to produce highly resistant **endospores (spores).** The spores are found in the soil, air, and all environments of the body. Species of *Bacillus* grow aerobically, and *Bacillus anthracis* is the cause of anthrax. *Clostridium* species grow anaerobically, and different species cause tetanus, botulism, and gas gangrene.

Bacillus and *Clostridium* species are also used for industrial purposes. *Bacillus thuringiensis* forms an insecticide useful against various forms of caterpillars, and *Clostridium* species are used to produce various types of chemicals, such as butanol.

Corynebacteria. **Corynebacteria** are pleomorphic members of the genus *Corynebacterium*, which are Gram-positive rods found in various environments, including the soil. The bacteria contain cytoplasmic phosphate granules that stain as characteristic **metachromatic granules.** One species, *Corynebacterium diphtheriae,* causes human diphtheria.

Actinomyces* and *Arthrobacter. *Actinomyces* species are Gram-positive rods that assume many shapes and usually form branching filaments. Most species are anaerobic, and one species is responsible for a human and cattle infection called lumpy jaw.

Arthrobacter species live primarily in the soil. These Gram-positive rods assume many shapes during their life cycles, including branching rods and spherical forms. *Arthrobacter* species are widely found in the soil, and many degrade herbicides and pesticides.

Acid-Fast Bacilli

Mycobacteria. The **mycobacteria** include species in the genus *Mycobacterium*. This group of rod-shaped bacteria possesses large amounts of mycolic acid in the cell wall. The presence of mycolic acid makes the bacteria very difficult to stain, but when heat or other agents are used to force carbolfuchsin into the cytoplasm, the bacteria resist decolorization with a dilute acid-alcohol solution. Therefore, they are said to be **acid-fast.**

Many mycobacteria are free-living, but two notable pathogens exist in the group: *M. tuberculosis,* the cause of tuberculosis in humans and cattle; and *M. leprae,* which causes leprosy.

Nocardioforms. **Nocardioforms** include nine genera of aerobic, acid-fast rods, including members of the genus *Nocardia.* Nocardioforms have aerial hyphae which project above the surface of their growth medium as branching filaments. The hyphae fragment into rods and cocci. Nocardioforms are found throughout nature in many types of soil and aquatic environments. One species, *N. asteroides,* causes infection of the human lung.

Archaebacteria

Archaebacteria differ from all other bacteria (which are sometimes called eubacteria). Archaebacteria are so named because biochemical evidence indicates that they evolved before the eubacteria and have not undergone significant change since then. The archaebacteria generally grow in extreme environments and have unusual lipids in their cell membranes and distinctive RNA molecules in their cytoplasm.

One group of archaebacteria are the **methanogens,** anaerobic bacteria found in swamps, sewage, and other areas of decomposing matter. The methanogens reduce carbon dioxide to methane gas in their metabolism. A second group are the **halobacteria,** a group of rods that live in high-salt environments. These bacteria have the ability to obtain energy from light by a mechanism different from the usual process of photosynthesis. The third type of archaebacteria are the **extreme thermophiles.** These bacteria live at extremely high temperatures, such as in hot springs, and are associated with extreme acid environments. Like the other archaebacteria, the extreme thermophiles lack peptidoglycan in their cell walls. Many depend on sulfur in their metabolism, and many produce sulfuric acid as an end-product.

Submicroscopic Bacteria

Rickettsiae. **Rickettsiae** are rod-shaped and coccoid bacteria belonging to the order Rickettsiales. These bacteria cannot be seen with the light microscope, and therefore the Gram stain is not used for identification. However, their walls have the characteristics of Gram-negative cell walls. Rickettsiae are obligate intracellular parasites that infect humans as well as arthropods such as ticks, mites, and lice. They are cultivated only with great difficulty in the laboratory and generally do not grow on cell-free media. Tissue cultures and fertilized eggs are used instead.

Rickettsiae are very important as human pathogens. Various species cause Rocky Mountain spotted fever, epidemic typhus, endemic typhus, scrub typhus, Q fever, and ehrlichiosis.

Chlamydiae. **Chlamydiae** are extremely tiny bacteria, below the resolving power of the light microscope. Although the Gram stain is not used for identification, the bacteria have cell walls resembling those in Gram-negative bacteria.

Chlamydiae display a growth cycle that takes place within host cells. The bacteria invade the cells and differentiate into dense bodies called **reticulate bodies.** The reticulate bodies reproduce and eventually form new chlamydiae in the host cell called **elementary bodies.** Chlamydiae cause several diseases in humans, such as psittacosis, a disease of the lung tissues; trachoma, a disease of the eye; and chlamydia, an infection of the reproductive tract.

Mycoplasmas. **Mycoplasmas** are extremely small bacteria, below the resolving power of the light microscope. They lack cell walls and are surrounded by only an outer plasma membrane. Without the rigid cell wall, the mycoplasmas vary in shape and are said to be **pleomorphic.** Certain species cause a type of mild pneumonia in humans as well as respiratory tract and urinary tract diseases.

Viruses are noncellular genetic elements that use a living cell for their replication and have an extracellular state. Viruses are ultramicroscopic particles containing nucleic acid surrounded by protein, and in some cases, other macromolecular components such as a membranelike envelope.

Outside the host cell, the virus particle is also known as a **virion.** The virion is metabolically inert and does not grow or carry on respiratory or biosynthetic functions.

At present, there are no technical names for viruses. International committees have recommended genus and family names for certain viruses, but the process is still in a developmental stage.

Viral Structure and Replication

Viruses vary considerably in size and shape. The smallest viruses are about 0.02 μm (20 nanometers), while the large viruses measure about 0.3 μm (300 nanometers). Smallpox viruses are among the largest viruses; polio viruses are among the smallest.

Viral structure. Certain viruses contain ribonucleic acid (RNA), while other viruses have deoxyribonucleic acid (DNA). The nucleic acid portion of the viruses is known as the **genome.** The nucleic acid may be single-stranded or double-stranded; it may be linear or a closed loop; it may be continuous or occur in segments.

The genome of the virus is surrounded by a protein coat known as a **capsid,** which is formed from a number of individual protein molecules called **capsomeres.** Capsomeres are arranged in a precise and highly repetitive pattern around the nucleic acid. A single type of capsomere or several chemically distinct types may make up the capsid. The combination of genome and capsid is called the viral **nucleocapsid.**

A number of kinds of viruses contain **envelopes.** An envelope is a membranelike structure that encloses the nucleocapsid and is obtained from a host cell during the replication process. The envelope contains viral-specified proteins that make it unique. Among the envelope viruses are those of herpes simplex, chickenpox, and infectious mononucleosis.

The nucleocapsids of viruses are constructed according to certain symmetrical patterns. The virus that causes tobacco mosaic disease, for example, has **helical symmetry.** In this case, the nucleocapsid is wound like a tightly coiled spiral. The rabies virus also has helical symmetry. Other viruses take the shape of an icosahedron, and they are said to have **icosahedral symmetry.** In an icosahedron, the capsid is composed of 20 faces, each shaped as an equilateral triangle (Figure 22). Among the icosahedral viruses are those that cause yellow fever, polio, and head colds.

The envelope of certain viruses is a lipid bilayer containing glycoproteins embedded in the lipid. The envelope gives a somewhat circular appearance to the virus and does not contribute to the symmetry of the nucleocapsid. Projections from the envelope are known as **spikes.** The spikes sometimes contain essential elements for attachment of the virus to the host cell. The virus of AIDS, the human immunodeficiency virus, uses its spikes for this purpose.

Bacteriophages are viruses that multiply within bacteria. These viruses are among the more complex viruses. They often have icosahedral heads and helical tails. The virus that attacks and replicates in *Escherichia coli* has 20 different proteins in its helical tail and a set of numerous fibers and "pins." Bacteriophages contain DNA and are important tools for viral research.

Viral replication. During the process of **viral replication,** a virus induces a living host cell to synthesize the essential components for the synthesis of new viral particles. The particles are then assembled into the correct structure, and the newly formed virions escape from the cell to infect other cells.

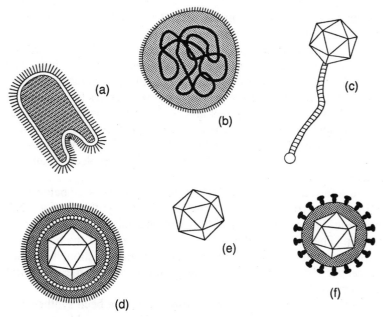

An array of viruses. (a) The helical virus of rabies. (b) The segmented helical virus of influenza. (c) A bacteriophage with an icosahedral head and helical tail. (d) An enveloped icosahedral herpes simplex virus. (e) The unenveloped polio virus. (f) The icosahedral human immunodeficiency virus with spikes on its envelope.

■ Figure 22 ■

The first step in the replication process is **attachment.** In this step, the virus adsorbs to a susceptible host cell. High specificity exists between virus and cell, and the envelope spikes may unite with cell surface receptors. Receptors may exist on bacterial pili or flagella or on the host cell membrane.

The next step is **penetration** of the virus or the viral genome into the cell. This step may occur by phagocytosis; or the envelope of the virus may blend with the cell membrane; or the virus may "inject" its genome into the host cell. The latter situation occurs with the bacteriophage when the tail of the phage unites with the bacterial cell wall and enzymes open a hole in the wall. The DNA of the phage penetrates through this hole.

The **replication** steps of the process occur next. The protein capsid is stripped away from the genome, and the genome is freed in the cell cytoplasm. If the genome consists of RNA, the genome acts as a messenger RNA molecule and provides the genetic codes for the synthesis of enzymes. The enzymes are used for the synthesis of viral genomes and capsomeres and the assembly of these components into new viruses. If the viral genome consists of DNA, it provides the genetic code for the synthesis of messenger RNA molecules, and the process proceeds.

In some cases, such as in HIV infection (as discussed below), the RNA of the virus serves as a template for the synthesis of a DNA molecule. The enzyme reverse transcriptase catalyzes the DNA's production. The DNA molecule then remains as part of the host cell's chromosome for an unspecified period. From this location, it encodes messenger RNA molecules for the synthesis of enzymes and viral components.

Once the viral genomes and capsomeres have been synthesized, they are assembled to form new virions. This **assembly** may take place in the cytoplasm or in the nucleus of the host cell. After the assembly is complete, the virions are ready to be released into the environment (Figure 23).

For the **release** of new viral particles, any of a number of processes may occur. For example, the host cell may be "biochemically exhausted," and it may disintegrate, thereby releasing the virions. For enveloped viruses, the nucleocapsids move toward the membrane of the host cell, where they force themselves through that membrane in a process called **budding**. During budding, a portion of cell membrane pinches off and surrounds the nucleocapsid as an envelope. The replication process in which the host cell experiences death is called the **lytic cycle** of reproduction. The viruses so produced are free to infect and replicate in other host cells in the area.

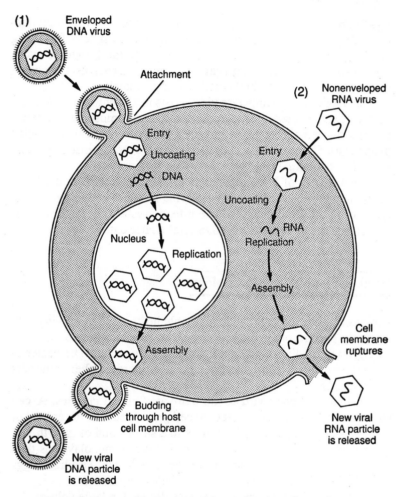

A generalized representation of the replication of two viruses. Replication of a DNA virus is shown in (1); replication of an RNA virus is displayed in (2).

■ Figure 23 ■

Lysogeny. Not all viruses multiply by the lytic cycle of reproduction. Certain viruses remain active within their host cells for a long period without replicating. This cycle is called the **lysogenic cycle.** The viruses are called **temperate viruses,** or **proviruses,** because they do not bring death to the host cell immediately.

In lysogeny, the temperate virus exists in a latent form within the host cell and is usually integrated into the chromosome. Bacteriophages that remain latent within their bacterial host cell are called **prophages.** This process is a key element in the recombination process known as **transduction.**

An example of lysogeny occurs in **HIV infection.** In this case, the human immunodeficiency virus remains latent within the host T-lymphocyte. An individual whose infection is at this stage will not experience the symptoms of AIDS until a later date.

Viral Cultivation and Physiology

Viruses can be cultivated within suitable hosts, such as a living cell. To study bacteriophages, for example, bacteria are grown in a suitable growth medium; then bacteriophages are added. The bacteriophages multiply within the bacteria and increase their numbers substantially.

Animal and plant viruses are cultivated in cell cultures. A **cell culture** is prepared by encouraging cell growth outside the animal or plant source. The cells are kept alive in a suspension of growth factors within a Petri dish. A thin layer of cells, or monolayer, is then inoculated with viruses, and replication takes place. Fertilized eggs and living animals can also be used to cultivate viruses.

For research study, viruses can be cultivated in large volumes by inoculations to tissue culture systems. After a time, the cells are degenerated, and viruses are harvested. The viral particles are concentrated by precipitation methods and purified by repeated centrifugations. Highly purified viruses can be obtained by crystallization and concentration under established conditions.

Viral measurements. Viruses are generally too small to be seen under the light microscope, and an electron microscope is usually necessary to make them visible. Although viruses can be quantified by observation, it is also possible to determine their number in terms of **virus infectious units,** each of which is the smallest unit that causes a detectable effect when viruses infect a susceptible host. Virus infectious units are expressed per volume of fluid.

One method for determining virus infectious units is by the **plaque assay.** The plaque assay is performed by cultivating viruses on a "lawn" of host cells and noting the presence of clear areas where viruses have replicated and destroyed the cells.

Another way of determining virus infectious units is by cultivating viruses in living animals and determining which dilution of virus is lethal to the animals. The **end-point dilution** can be calculated by this method.

Antiviral agents. The antibiotics normally used to treat bacterial disease cannot be used to inactivate viruses because viruses do not perform the biochemical functions that antibiotics interfere with. For example, penicillin is used to interrupt the synthesis of the bacterial cell wall, but viruses have no cell wall.

However, there are several nucleotide analog drugs that interfere with viral replication. **Acyclovir,** for example, is used against herpes viruses because this drug prevents the synthesis of DNA during viral replication. A drug called **azidothymidine (AZT)** is used for patients with HIV infection because this drug also prevents the synthesis of DNA. A drug called **ganciclovir** is used against cytomegaloviruses, and **amantadine** is useful against influenza viruses.

Interferon, a naturally produced antiviral agent approved for certain uses, is a group of proteins produced by host cells after they have been infected by viruses. The interferons do not protect the host cell, but they do provide protection to neighboring cells against viral replication. Interferons can be produced by genetic engineering methods.

Viral vaccines. Protection against viral disease can be rendered by using a **viral vaccine.** Viral vaccines can be composed of inactivated or attenuated viruses. **Inactivated viruses** ("dead viruses") are unable to replicate in host cells because of some chemical or physical treatment. The Salk vaccine against polio and the yellow fever vaccine are examples.

Attenuated viruses ("live viruses") are weakened viruses that replicate at a very slow rate in host cells and generally do not produce any symptoms of disease when inoculated to humans. Attenuated viruses are used in the Sabin polio vaccine and in the vaccines against measles and rubella. The most contemporary vaccines are composed of viral proteins produced by **genetic engineering methods.** The vaccine for hepatitis B is an example of this type of vaccine.

Viral inactivation. Virus particles are composed of nucleic acid, protein, and in some cases, a lipid envelope. As such, the viruses are susceptible to normal inactivation by chemical substances that react with any of these organic compounds. Such things as chlorine, iodine, phenol, detergents, and heavy metals rapidly inactivate viruses. In addition, viruses are destroyed by heating methods used for other microorganisms, and they are very susceptible to the effects of ultraviolet light. Filters can be used to remove viruses from fluids so long as the filter pores are small enough to trap viral particles.

The **fungi** (singular, **fungus**) include several thousand species of eukaryotic, sporebearing organisms that obtain simple organic compounds by absorption. The organisms have no chlorophyll and reproduce by both sexual and asexual means. The fungi are usually filamentous, and their cell walls have **chitin.** The study of fungi is called **mycology,** and fungal diseases are called **mycoses.**

Together with bacteria, fungi are the major decomposers of organic materials in the soil. They degrade complex organic matter into simple organic and inorganic compounds. In doing so, they help recycle carbon, nitrogen, phosphorous, and other elements for reuse by other organisms. Fungi also cause many plant diseases and several human diseases.

Two major groups of organisms make up the fungi. The filamentous fungi are called molds, while the unicellular fungi are called yeasts. The fungi are classified in the kingdom Fungi in the Whittaker five-kingdom system of classification.

Structure and Physiology

There is considerable variation in the structure, size, and complexity of various fungal species. For example, fungi include the microscopic yeasts, the molds seen on contaminated bread, and the common mushrooms.

Molds consist of long, branching filaments of cells called **hyphae** (singular, **hypha**). A tangled mass of hyphae visible to the unaided eye is a **mycelium** (plural, **mycelia**). In some molds, the cytoplasm passes through and among cells of the hypha uninterrupted by cross walls. These fungi are said to be **coenocytic fungi.** Those fungi that have cross walls are called **septate fungi,** since the cross walls are called septa.

Yeasts are microscopic, unicellular fungi with a single nucleus and eukaryotic organelles. They reproduce asexually by a process of

budding. In this process, a new cell forms at the surface of the original cell, enlarges, and then breaks free to assume an independent existence.

Some species of fungi have the ability to shift from the yeast form to the mold form and vice versa. These fungi are **dimorphic.** Many fungal pathogens exist in the body in the yeast form but revert to the mold form in the laboratory when cultivated.

Reproduction in yeasts usually involves **spores.** Spores are produced by either sexual or asexual means. Asexual spores may be free and unprotected at the tips of hyphae, where they are called **conidia** (Figure 24). Asexual spores may also be formed within a sac, in which case they are called **sporangiospores.**

Nutrition. Fungi grow best where there is a rich supply of organic matter. Most fungi are saprobic (obtaining nutrients from dead organic matter). Since they lack photosynthetic pigments, fungi cannot perform photosynthesis and must obtain their nutrients from preformed organic matter. They are therefore **chemoheterotrophic organisms.**

Most fungi grow at an acidic pH of about 5.0, although some species grow at lower and higher pH levels. Most fungi grow at about 25°C (room temperature) except for pathogens, which grow at 37°C (body temperature). Fungi store glycogen for their energy needs and use glucose and maltose for immediate energy metabolism. Most species are aerobic, except for the fermentation yeasts that grow in both aerobic and anaerobic environments.

Reproduction. **Asexual reproduction** occurs in the fungi when spores form by mitosis. These spores can be conidia, sporangiospores, arthrospores (fragments of hyphae), or chlamydospores (spores with thick walls).

During **sexual reproduction,** compatible nuclei unite within the mycelium and form sexual spores. Sexually opposite cells may unite within a single mycelium, or different mycelia may be required. When the cells unite, the nuclei fuse and form a diploid nucleus. Several divisions follow, and the haploid state is reestablished.

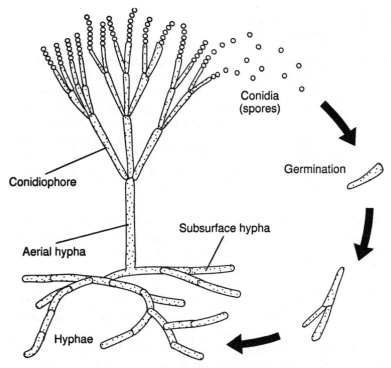

Conidia
(spores)

Germination

Conidiophore

Subsurface hypha

Aerial hypha

Hyphae

The microscopic structures of a septate fungus showing asexually produced conidia that leave the fungus and germinate to produce a new mycelium.

■ Figure 24 ■

Fungal spores are important in the identification of the fungus, since the spores are unique in shape, color, and size. A single spore is capable of germinating and reestablishing the entire mycelium. Spores are also the method for spreading fungi in the environment. Finally, the nature of the sexual spores is used for classifying fungi into numerous divisions.

Classification of Fungi

Division Zygomycota. Members of the division **Zygomycota** are known as **zygomycetes.** Zygomycetes produce sexual spores known as **zygospores** (Figure 25), as well as asexual sporangiospores.

A familiar member of the division is *Rhizopus stolonifer,* a fungus found on fruits, vegetables, and breads. It is the familiar bread mold. It anchors itself to the substratum with special hyphae known as **rhizoids.** *Rhizopus* is used in the industrial production of steroids, meat tenderizers, industrial chemicals, and certain coloring agents.

Division Ascomycota. Members of the division **Ascomycota** are referred to as **ascomycetes.** After sexual fusion of cells has taken place, these organisms form their sexual spores within a sac called an **ascus.** Therefore, they are called **sac fungi.**

Ascomycetes include the powdery mildews and the fungi that cause Dutch elm disease and chestnut blight disease. The research organism *Neurospora crassa* is found within this group. Asexual reproduction in the ascomycetes involves conidia.

Many yeasts are classified in the division Ascomycota. Of particular interest is the **fermentation yeast** *Saccharomyces.* This yeast is used in the production of alcoholic drinks, in bread making, and as a source of growth factors in yeast tablets. It is an extremely important research organism as well.

Division Basidiomycota. Members of the division **Basidiomycota** are referred to as **basidiomycetes** and are called **club fungi.** After the sexual cells have united, they undergo division and produce a club-shaped structure called a **basidium.** Sexually produced **basidiospores** form at the tips of the basidia. Basidia are often found on huge, visible, fruiting bodies called **basidiocarps.** The typical **mushroom** is a basidiocarp.

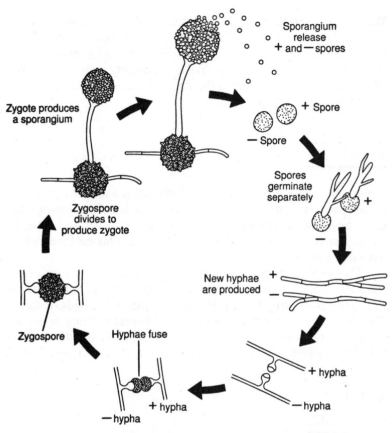

Sexual reproduction in the mold Rhizopus stolonifer. *Plus and minus mycelia produce sexually opposite hyphae that fuse and give rise to zygospores, which germinate to form new mycelia.*

■ Figure 25 ■

Basidiomycetes are used as food (for example, mushrooms), but some basidiomycetes are pathogens. One of the organisms of meningitis is the basidiomycete *Cryptococcus neoformans*. The mushroom *Amanita* is poisonous to humans.

Division Deuteromycota. Members of the **Deuteromycota** division are called **deuteromycetes.** These fungi lack a known sexual cycle of reproduction and are said to be "imperfect." When its sexual cycle is discovered, a fungus from this division is usually reclassified in one of the other divisions. Among the imperfect fungi are the organisms of athlete's foot and ringworm.

Slime Molds

Slime molds have characteristics of both molds and protozoa. Under certain conditions, the slime mold exists as masses of cytoplasm, similar to amoebae. It moves over rotting logs or leaves and feeds by phagocytosis. The amoeba stage is called the **plasmodium,** which has many nuclei.

The amoeba stage ends when the plasmodium matures or encounters a harsh environment. At this point, it moves to a light area and develops fruiting bodies that form spores at the ends of stalks. The spores are resistant to environmental excesses. They germinate when conditions are suitable to form flagellated **swarm cells,** or amoeboid cells, which later fuse to again form a multinucleate plasmodium.

Water Molds

Water molds belong to the group known as **oomycetes.** The water molds resemble other fungi because they have branched filaments and form spores. However, the water molds have cellulose in their cell walls, while other fungi have chitin.

Oomycetes have a complex reproductive cycle which includes flagella-bearing **zoospores.** Certain water molds are parasites of fish. Others cause disease in plants such as tobacco, grapes, and potatoes.

Algae are eukaryotic organisms that have no roots, stems, or leaves but do have chlorophyll and other pigments for carrying out photosynthesis. Algae can be multicellular or unicellular.

General Characteristics

Unicellular algae occur most frequently in water, especially in plankton. **Phytoplankton** is the population of free-floating microorganisms composed primarily of unicellular algae. In addition, algae may occur in moist soil or on the surface of moist rocks and wood. Algae live with fungi in **lichens.**

According to the Whittaker scheme, algae are classified in seven divisions, of which five are considered to be in the Protista kingdom and two in the Plantae kingdom. The cell of an alga has eukaryotic properties, and some species have flagella with the "9-plus-2" pattern of microtubules. A nucleus is present, and multiple chromosomes are observed in mitosis. The chlorophyll and other pigments occur in **chloroplasts,** which contain membranes known as **thylakoids.**

Most algae are **photoautotrophic** and carry on photosynthesis. Some forms, however, are **chemoheterotrophic** and obtain energy from chemical reactions and nutrients from preformed organic matter. Most species are saprobes, and some are parasites.

Reproduction in algae occurs in both asexual and sexual forms. Asexual reproduction occurs through the fragmentation of colonial and filamentous algae or by spore formation (as in fungi). Spore formation takes place by mitosis. Binary fission also takes place (as in bacteria).

During sexual reproduction, algae form differentiated sex cells that fuse to produce a diploid **zygote** with two sets of chromosomes. The zygote develops into a sexual spore, which germinates when conditions are favorable to reproduce and reform the haploid organism

having a single set of chromosomes. This pattern of reproduction is called **alternation of generations.**

Divisions of Unicellular Algae

Five divisions of unicellular algae are considered in microbiology because of their microscopic form and their unicellular characteristic. These organisms are classified in the kingdom Protista.

Division Chlorophyta. Algae of the division **Chlorophyta** possess green chlorophyll pigments and carotenoid pigments. A representative member is *Chlamydomonas,* which is often used in research and as a laboratory specimen. *Chlamydomonas* produces **zoospores,** which are flagellated. Organisms such as *Chlamydomonas* are believed to be evolutionary ancestors of other species. Other organisms in the division are *Volvox* and *Spirogyra.*

Division Charophyta. Members of the division **Charophyta** are **stoneworts.** Stoneworts cover the bottoms of ponds and may be a source of limestone.

Division Euglenophyta. Members of the division **Euglenophyta** include the common organism *Euglena.* These organisms have chlorophyll and carotenoid pigments for photosynthesis and flagella for movement. They share many characteristics with both plants and animals and are believed to be a basic stock of evolution.

A typical *Euglena* cell has a large nucleus and nucleolus. Contractile vacuoles help empty water from the organism, and two flagella arise at one end of the cell. Reproduction occurs by binary fission in the longitudinal plane.

Division Chrysophyta. Members of the division **Chrysophyta** are **brown** and **yellow-green algae.** These organisms contain chlorophyll pigments as well as special carotenoid pigments called fucoxanthins. Fucoxanthins give the golden-brown color to members of the division. Members of the division include the **diatoms,** oceanic photosynthetic algae found at the bases of many food chains. Diatoms contribute immense amounts of oxygen to the atmosphere and occupy key places in the spectrum of living things because they convert the sun's energy into the energy in carbohydrates.

Division Pyrrophyta. Members of the division **Pyrrophyta** are pigmented marine forms that include the **dinoflagellates,** amoeboid cells with flagella as well as protective cellulose plates that surround the cells. They have chlorophyll, carotenoid, and xanthophyll pigments. Dinoflagellates often have a brown or yellow color, and they reproduce by longitudinal division through mitosis. Dinoflagellates make up a large portion of marine plankton and are essential to many of the ocean food chains. Certain species are luminescent. Others have red or orange pigments; when these organisms multiply at abnormally high rates, they cause the "red tides."

Protozoa are eukaryotic microorganisms. Although they are often studied in zoology courses, they are considered part of the microbial world because they are unicellular and microscopic.

General Characteristics

Protozoa are notable for their ability to move independently, a characteristic found in the majority of species. They usually lack the capability for photosynthesis, although the genus *Euglena* is renowned for motility as well as photosynthesis (and is therefore considered both an alga and a protozoan). Although most protozoa reproduce by asexual methods, sexual reproduction has been observed in several species. Most protozoal species are aerobic, but some anaerobic species have been found in the human intestine and animal rumen.

Protozoa are located in most moist habitats. Free-living species inhabit freshwater and marine environments, and terrestrial species inhabit decaying organic matter. Some species are parasites of plants and animals.

Protozoa play an important role as **zooplankton,** the free-floating aquatic organisms of the oceans. Here, they are found at the bases of many food chains, and they participate in many food webs.

Size and shape. Protozoa vary substantially in size and shape. Smaller species may be the size of fungal cells; larger species may be visible to the unaided eye. Protozoal cells have no cell walls and therefore can assume an infinite variety of shapes. Some genera have cells surrounded by hard shells, while the cells of other genera are enclosed only in a cell membrane.

Many protozoa alternate between a free-living vegetative form known as a **trophozoite** and a resting form called a **cyst.** The protozoal cyst is somewhat analogous to the bacterial spore, since it resists harsh conditions in the environment. Many protozoal parasites are taken into the body in the cyst form.

Most protozoa have a single nucleus, but some have both a macronucleus and one or more micronuclei. Contractile vacuoles may be present in protozoa to remove excess water, and food vacuoles are often observed.

Nutrition and locomotion. Protozoa are **heterotrophic** microorganisms, and most species obtain large food particles by **phagocytosis.** The food particle is ingested into a food vacuole. Lysosomal enzymes then digest the nutrients in the particle, and the products of digestion are distributed throughout the cell. Some species have specialized structures called **cytostomes,** through which particles pass in phagocytosis.

Many protozoal species move independently by one of three types of locomotor organelles: flagella, cilia, and pseudopodia. **Flagella** and **cilia** are structurally similar, having a "9-plus-2" system of microtubules, the same type of structure found in the tail of animal sperm cells and certain cells of unicellular algae. How a protozoan moves is an important consideration in assigning it to a group.

Classification of Protozoa

All protozoal species are assigned to the kingdom **Protista** in the Whittaker classification. The protozoa are then placed into various groups primarily on the basis of how they move. The groups are called phyla (singular, phylum) by some microbiologists, and classes by others. Members of the four major groups are illustrated in Figure 26.

Mastigophora. The **Mastigophora** are protozoa having one or more **flagella.** Reproduction in these protozoa generally occurs by fission, although sexual reproduction is observed in some species. Members of the group include the organisms that cause intestinal distress

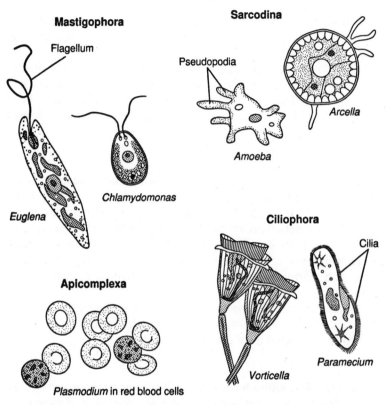

Mastigophora

Flagellum

Euglena

Chlamydomonas

Sarcodina

Pseudopodia

Arcella

Amoeba

Apicomplexa

Plasmodium in red blood cells

Ciliophora

Cilia

Vorticella

Paramecium

An array of protozoa showing representatives of the four major groups.

■ Figure 26 ■

(*Giardia lamblia*) and sleeping sickness (*Trypanosoma brucei*) in humans. Some of these protozoa have only a single flagellum (for example, *Trypanosoma*), but others may have up to eight (for example, *Giardia*). Flagellar motion tends to be jerky, so flagellates can be identified in live specimens by this characteristic movement.

Also in this group is *Euglena gracilis,* a flagellated organism that also has chlorophyll pigments for photosynthesis. For this reason, the organism is considered both a unicellular alga and a protozoan. Since it has the characteristics of both animals (motion) and plants (photosynthesis), many evolutionists consider *Euglena* to be a source of all complex organisms.

Flagellates are very common in nature, and numerous species may be observed in pond and swamp water, aquarium water, and birdbath water. The flagellum may either push the protozoan ahead or lead the cell along. A complex basal structure at one end of the cell anchors the flagellum to the cell.

Sarcodina. Members of the group **Sarcodina** are amoebas. These organisms move by **pseudopodia,** although flagella may be present in the reproductive stages. Cytoplasmic streaming assists movement. Asexual reproduction occurs by fission of the cell. Sarcodina includes two marine groups known as foraminiferans and radiolarians. Both groups were present on earth when the oil fields were in formative stages, and marine geologists use them as potential markers for oil fields.

Another member, *Entamoeba histolytica,* is the cause of amoebic dysentery. This organism can cause painful lesions of the intestine if it is contracted from contaminated water.

Ciliophora. The **Ciliophora** group includes protozoa that move by means of **cilia.** Two types of nuclei, the macronucleus and the micronucleus, are often found in the cells, and a contractile vacuole is usually present. Conjugation may be used for sexual reproduction, and binary fission also occurs. The distinctive rows of cilia vibrate in synchrony and propel the organism in one direction.

Paramecium is a member of this group, as are many other free-living ciliates such as *Tetrahymena. Paramecium* cells are known for their ability to conjugate and exchange genetic material, thereby acquiring new genetic characteristics. The cells also have **kappa factors,** which are chemical substances known to destroy other protozoal cells, and they produce **trichocysts,** which are submicroscopic stinging particles. Indeed, the physiology of a single *Paramecium* cell is similar to that of a multicellular organism.

Apicomplexa. Members of the group **Apicomplexa** form **spores** at one stage in their life cycle. For this reason, the group is also known as **Sporozoa.** Reproduction is a complex phenomenon in this group, and all species are parasites. Members of the group display no means of locomotion in the adult form. The agents of malaria and toxoplasmosis are classified here. Another important pathogen is *Pneumocystis carinii,* a cause of sometimes-lethal pneumonia in AIDS patients; over half the deaths known to be associated with AIDS have been caused by *P. carinii.*

A delicate relationship exists between pathogenic microorganisms and body defenses. When the defenses resist the pathogens, the body remains healthy. But when the pathogens overcome the defenses, the result is disease. Once disease has been established, the infected individual may suffer temporary or permanent damage or may experience death. The outcome depends upon many factors attending the episode of disease.

Infection and Disease

The scientific study of disease is called **pathology,** from the Greek "pathos" meaning suffering. Pathology is concerned with the cause of disease, called the **etiology** (the agent of disease is the **etiologic agent**). It also deals with **pathogenesis,** the manner in which a disease develops. Pathology is also concerned with the structural and functional changes brought about by the disease in tissues.

The terms infection and disease do not have identical meaning. **Infection** refers to an invasion of body tissues by microorganisms; **disease** is a change from the state of good health resulting from a microbial population living in the tissues (Figure 27). Infection may occur without disease. For example, the flora of microorganisms always present on the body's skin is a type of infection but not disease.

The normal flora. The **normal flora** is the population of microorganisms found where the body tissues interface with the environment. Much of the normal flora is permanent, but some portions are transient. The **transient flora** is present for a time and then disappears.

Various types of relationships exist between the normal flora and the body. The general name of a relationship is **symbiosis,** a term that means living together. In some cases, the symbiosis is further identi-

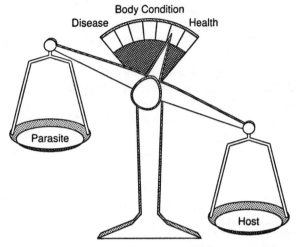

The balance between health and disease. The condition of the human body depends on interactions between host (the body) and parasite (the infectious microorganism). When the human body wins the battle for supremacy, the result is health and the rejection of disease.

■ Figure 27 ■

fied as a **commensalism,** when one organism benefits and the other remains unaffected. A type of symbiosis called **mutualism** exists when both organisms benefit one another. A symbiosis called **parasitism** develops when one organism damages the other.

Opportunistic organisms. Certain organisms of the normal flora are opportunistic. **Opportunistic organisms** are potentially pathogenic organisms that normally do not cause disease. However, in a compromised host, the organisms may see "opportunity" to invade the tissues. An example occurs in individuals who have AIDS. Opportunistic organisms such as *Pneumocystis carinii* invade the lung tissues and cause a lethal pneumonia.

The Development of Infectious Disease

Infectious diseases are those caused by microorganisms. In order to relate a particular organism to a particular disease, **Koch's postulates** must be fulfilled. First devised by Robert Koch in the 1870s, Koch's postulates are a series of procedures for identifying the cause of a particular disease. They are described in the first chapter of this book.

Symptoms and signs. Infectious diseases are usually characterized by changes in body function known as **symptoms.** Symptoms are subjective changes not always apparent to the observer. The patient may also exhibit **signs,** which are objective changes that can be measured. Fever and a skin rash are examples of signs. When a specific group of symptoms or signs accompanies a disease, the group is called a **syndrome.**

Transmission and incidence. Infectious diseases may be classified according to their **transmissibility.** A disease that spreads from one host to another is a **communicable disease.** Those communicable diseases transmitted with particular ease are said to be **contagious.** Diseases not spread between hosts are **noncommunicable.** Staphylococcal food poisoning is an example.

The **incidence** of a disease refers to the percent of a population that contracts it over a particular period. The **prevalence** of a disease, by contrast, is the percentage of a population having the disease at a particular time.

When a disease occurs only occasionally, it is called a **sporadic disease.** A disease present in a population at all times is an **endemic disease.** A disease that breaks out in a population in a short period is an **epidemic disease,** and an epidemic disease occurring throughout the world is a **pandemic disease.**

Types of disease. Diseases can be defined in terms of their severity and duration. An **acute disease** occurs rapidly and lasts a short time, while a **chronic disease** develops slowly and lasts a long time. Influenza is an acute disease, while tuberculosis is a chronic disease. A **subacute disease** is a disease that has vague symptoms and lasts a relatively long time. A **latent disease** remains inactive in a host for a time and then becomes active.

Infections can be described as **local infections** if they are restricted to a small area of the body and **systemic infections** if they spread throughout the body systems. The presence of multiplying microorganisms in the blood is **septicemia.** Toxins present in the blood constitute **toxemia.**

Infectious diseases may also be described as primary diseases or secondary diseases. A **primary disease** is the first illness that occurs, and a **secondary disease** is due to an opportunistic microorganism, often a normal resident, after the body's defenses have weakened.

Modes of disease transmission. When a disease remains in a population, a source of pathogens called a **reservoir of infection** exists within the population. The reservoir can be human, animal, or nonliving, such as the soil. A human reservoir who has had the disease and recovered but continues to shed infectious organisms is called a **carrier.** Animal diseases spread to humans are called **zoonoses.**

Among the principal routes of transmission of disease are contact, vectors, and vehicles. **Contact** can be direct or indirect. **Direct transmission** occurs from person to person by such things as touching, kissing, and sexual intercourse (Figure 28). **Indirect transmission** occurs when a nonliving object is intermediary between two humans.

A lifeless object known as a **fomite** is often involved in disease transmission. The object may be a towel, cup, or eating utensil. Transmission can also be effected by **droplet nuclei,** bits of mucus and saliva that spread between individuals.

Vectors are living things. **Arthropods** such as mosquitoes, flies, and ticks may carry pathogens on their body parts, in which case they are **mechanical vectors.** If the arthropod is infected and transmits the organism in its saliva or feces, it is a **biological vector.**

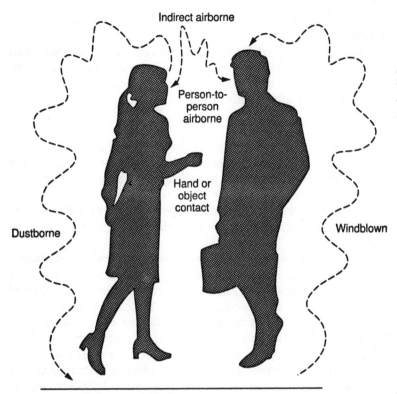

Indirect airborne

Person-to-person airborne

Hand or object contact

Dustborne

Windblown

Settles in dust

Some modes of transmission of microorganisms from the respiratory and oral tract.

■ Figure 28 ■

Vehicles are lifeless objects such as food, water, and air. Water may be contaminated by human feces, while food is often contaminated by pathogens from the soil. Air can be a vehicle for transmission for droplet nuclei in such diseases as tuberculosis and common colds.

In order for infection to be transmitted, microorganisms must leave the body through a **portal of exit,** which can be the intestine, mouth, or skin surface. Generally, the portal of exit is the same as the infected body part. Organisms enter the new individual through a **portal of entry.**

Infections acquired during a hospital stay are called **nosocomial infections.** These infections often occur in compromised hosts who are being treated for other conditions such as cancer, nutritional deficiency, burns, or other forms of stress.

Disease patterns. When a disease develops in an individual, a recognized set of periods can be identified. The first period is the **period of incubation,** the time between the entry of the parasite into the host and the appearance of symptoms.

The next period is the **prodrome period.** This period is accompanied by mild symptoms such as aches, fever, and early signs of disease.

Next comes the **period of illness,** when the disease is most acute. Signs and symptoms are most apparent, and each disease has its own characteristic appearance. The body's immune system is activated during this period, and specific defense is critical to recovery.

The final periods are the periods of decline and convalescence. The **period of decline** is one in which the signs and symptoms subside, and during the **period of convalescence,** the person returns to normal. After the disease has abated, the immune system continues to produce antimicrobial factors that will ensure long-term immunity.

Infectious disease results from a competition for supremacy between the parasite and the host. If the parasite overcomes the host, there is a change in the general state of good health and disease develops.

Several contributing factors are involved in the establishment of infectious disease. These factors determine whether the infecting organism will survive in the body.

Contributing Factors

Portals of entry. In order for a pathogen to gain access to the host, the pathogen must pass through a **portal of entry.** One of the most common portals of entry is the **mucous membranes,** especially those of the respiratory, gastrointestinal, and urogenital tracts. Another important portal of entry is the **skin.** Penetration of the skin occurs during a wound or by a hair follicle. When microorganisms penetrate below the skin, the portal of entry is said to be the **parenteral route.**

Dose. The **dose** of an organism refers to the number of microorganisms required to establish an infection. For some diseases, such as typhoid fever, the dose is a few hundred bacteria. For other diseases, such as cholera, the dose may be several million bacteria. The dose may be expressed as the LD_{50}, which refers to the dose of microorganisms that will kill 50 percent of the hosts it enters.

Invasiveness. **Invasiveness** is a property that encourages disease because it refers to the ability of pathogens to penetrate into the tissues. Those organisms that cause intestinal ulcers, such as *Entamoeba histolytica,* penetrate the tissue effectively. Tissue invasion often begins with **adherence,** the ability of pathogens to attach to the tissue by using structures such as pili. The presence of a capsule or glycocalyx encourages adherence because they are composed of sticky materials.

Capsules. Microorganisms that possess **capsules** are able to resist host defenses by interfering with phagocytosis. Normally, the body uses white blood cells to engulf and destroy pathogens. However, toxic substances in the capsule are able to destroy the white blood cells before the white blood cells perform phagocytosis. The organism of pneumonia *Streptococcus pneumoniae* is well known for the toxic materials in its capsule. Many other pathogens also possess capsules.

Enzymes and Toxins

Enzymes. Many pathogens produce a series of **enzymes** to help overcome body defenses and establish themselves in the host. One example is **leukocidins,** a group of enzymes that destroy white blood cells. This destruction lessens the body's ability to perform phagocytosis.

Other bacterial enzymes are **hemolysins.** These enzymes destroy red blood cells. Streptococci, staphylococci, and certain *Clostridium* species produce hemolysins.

Coagulases are bacterial enzymes that clot the blood. These enzymes convert fibrinogen into fibrin, which forms the threads of a blood clot. The clot helps staphylococci avoid the body's phagocytes and contributes to its pathogenicity.

Other important enzymes are streptokinase and hyaluronidase. **Streptokinase** is a streptococcal enzyme that dissolves blood clots. This activity helps the organism escape the body's attempt to wall off an infection. **Hyaluronidase** destroys hyaluronic acid, a polysaccharide that "cements" cells together in a tissue. Hyaluronidase thus permits organisms to spread through tissues and establish themselves at sites distant from that of the intial infection. Another enzyme, called

collagenase, breaks down collagen in the connective tissues of muscles. It thereby encourages the spread of infection.

Toxins. Many bacteria are able to produce poisonous substances called **toxins.** Toxins act on the body's cells, tissues, and organs and interfere with important body processes, thereby interrupting normal body functions. Those microorganisms that produce toxins are said to be **toxigenic.** The condition in which toxins are produced is called **toxemia.**

Two important types of toxins are exotoxins and endotoxins. **Exotoxins** are proteins produced by bacteria during their growth and liberated into their surrounding environment. Exotoxins are produced chiefly by Gram-positive bacteria, and the genes for this production are carried primarily on the plasmids.

Various types of exotoxins exist. **Neurotoxins** interfere with the nervous system, while **enterotoxins** interfere with activities of the gastrointestinal tract. In response to toxins, the body produces special antibodies called **antitoxins,** which unite with and neutralize the toxins, providing defense against disease.

It is possible to immunize against the effects of exotoxins by injecting **toxoids** into individuals. Toxoids are preparations of exotoxins chemically treated to destroy their toxigenicity but retain their ability to elicit antibody formation in the body. Toxoids are currently available to protect against diphtheria and tetanus (the DT injection).

Endotoxins are portions of the cell wall of Gram-negative bacteria. They consist primarily of lipopolysaccharides and are released when bacteria break apart during the process of lysis. Since lysis occurs during antibiotic therapy, the effects of endotoxins can bring about a worsening of symptoms during the recovery period. This condition is called **endotoxin shock.** It is accompanied by fever, chills, aches, and cardiovascular collapse.

Pathogenic Viruses

Because viruses lack metabolic capabilities, they rely on other means for overcoming body defenses and causing disease. Viruses avoid body defenses by multiplying within host cells, where antibodies and other components of the immune system cannot reach them.

The effect occurring in host cells during viral invasion is referred to as the **cytopathic effect.** The cytopathic effect can develop when the virus alters the metabolism of the cell and prevents it from producing essential cellular components. Alternatively, the virus may induce cells to cling together in a large mass called a **syncytium.** In some cases, the virus causes the cell's lysosomes to release enzymes which then destroy the cell.

The host-parasite relationship may result in an episode of disease. In this case, the body will mount two general forms of defense: nonspecific defense and specific defense centered in the immune system. This chapter is concerned with the nonspecific types of defense.

Nonspecific Mechanisms of Defense

The body possesses many mechanisms that impart nonspecific defense. The objectives of these mechanisms are to prevent microorganisms from gaining a foothold in the body and to destroy them if they penetrate to the deeper tissues.

Mechanical barriers. **Mechanical barriers** at the portal of entry represent the first line of defense for the body. These defenses are normally part of the body's anatomy and physiology. The **skin** is a representative example. The outermost layers of skin consist of compacted, cemented cells impregnated with the insoluble protein keratin. The thick top layer is impervious to infection and water. In the unbroken state, it usually is not penetrated by pathogens.

The **mucous membranes** of the urinary, respiratory, and digestive tracts are another example. They are moist and permeable, but their fluids, such as tears, mucus, and saliva, rid the membrane of irritants. **Nasal hairs** trap particles in the respiratory tract, and the fluids exert a flushing action. **Cilia** on the cells sweep and trap particles in the respiratory tract, and coughing ejects irritants.

Chemical defenses. Among the nonspecific chemical defenses of the body are the secretions of lubricating glands. The tears and saliva contain the enzyme **lysozyme,** which breaks down the peptidoglycan of the cell wall of Gram-positive bacteria. The **lactic acid** of the vagina imparts defense, and the extremely caustic **hydrochloric acid** of the stomach is a barrier to the intestine. Semen contains the antimicrobial substance **spermine** that inhibits bacteria in the male urogenital tract.

Genetic barriers. The hereditary characteristics of an individual are a deterrent to disease as well. For example, humans suffer HIV infection because their T-lymphocytes have the receptor sites for the human immunodeficiency virus. Dogs, cats, and other animals are immune to this disease because they do not possess the genes for producing the receptor sites. Conversely, humans do not suffer canine distemper because humans lack the appropriate receptor sites for the virus that causes the disease.

Inflammation. **Inflammation** is a nonspecific response to any trauma occurring to tissues. It is accompanied by signs and symptoms that include heat, swelling, redness, and pain. Inflammation mobilizes components of the immune system, sets into motion repair mechanisms, and encourages phagocytes to come to the area and destroy any microorganisms present.

Inflammation can be controlled by nervous stimulation and chemical substances called **cytokines.** These chemical products of tissue cells and blood cells are responsible for many of the actions of inflammation. The loss of fluid leads to a local swelling called **edema.** In some types of inflammation, phagocytes accumulate in the whitish mass of cells, bacteria, and debris called **pus.**

Fever. **Fever** is considered a nonspecific defense mechanism because it develops in response to numerous traumas. Fever is initiated by circulating substances called **pyrogens,** which affect the brain's hypothalamus and cause the latter to raise the temperature. Although excessive fever can be dangerous, fever is believed to have a beneficial role because it retards the growth of temperature-sensitive microorganisms (for example, leprosy bacilli), and it increases the metabolism of body cells while stimulating the immune reaction and the process of phagocytosis.

Interferon. **Interferon** is a group of antiviral substances produced by body cells in response to the presence of viruses. Lymphocytes and macrophages produce **alpha-interferon,** epithelial cells produce **beta-interferon,** and T-lymphocytes produce **gamma-interferon.** The interferons do not directly inhibit viruses. Instead, they stimulate adjacent cells to produce substances that inhibit the replication of viruses in those cells. Interferons produced in response to one virus will protect against many other types of viruses, and for this reason, interferon is considered a nonspecific form of defense.

Phagocytosis

Phagocytosis is a nonspecific defense mechanism in which various phagocytes engulf and destroy the microorganisms of disease.

Phagocytes. Among the important phagocytes are the circulating white blood cells called **neutrophils** and **monocytes.** In the tissues, the monocytes are transformed into phagocytic cells called **macrophages.** The macrophages move through the tissues of the body performing phagocytosis and destroying parasites. They are part of the **reticuloendothelial system.** Phagocytes also initiate the processes of the immune system.

The process of phagocytosis begins with **attachment** and **ingestion** of microbial particles (Figure 29) into a bubblelike organelle called a **phagosome.** Once inside the phagocyte, the phagosome containing the microorganism joins with a **lysosome,** which contributes enzymes. The fusion of phagosome and lysosome results in a **phagolysosome.** Microorganisms are destroyed within minutes, and

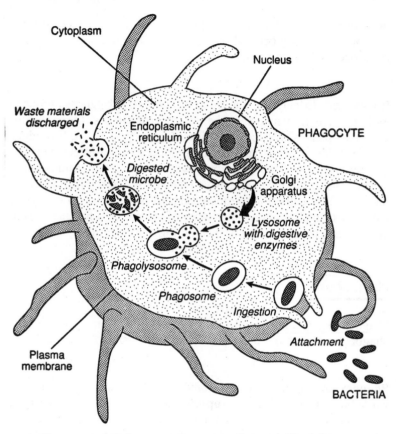

Cytoplasm

Nucleus

Waste materials
discharged

Endoplasmic
reticulum

PHAGOCYTE

Digested
microbe

Golgi
apparatus

Lysosome
with digestive
enzymes

Phagolysosome

Phagosome

Ingestion

Plasma
membrane

Attachment

BACTERIA

*The process of phagocytosis, a type of nonspecific defense to
disease.*

■ Figure 29 ■

the microbial debris is eliminated from the cell in the process of **eges-
tion.** In the immune process, chemical portions of the microorganism
called **antigenic determinants** are displayed on the surface of the
phagocyte to stimulate the immune process.

Phagocytosis is enhanced by products of the immune system
called **antibodies.** These protein molecules bind to microorganisms
and encourage engulfing by phagocytosis.

The complement system. The complement system is a series of proteins that circulate in the blood and encourage phagocytosis or otherwise "complete" the defensive process. Many immune reactions stimulate the complement system.

The complement system operates in a **cascade of reactions.** In the pathway, certain complement components react with one another and produce new substances that induce other components to react. The results of the myriad reactions are substances that induce other complement components into action. The overall result is a number of substances toxic to microorganisms. The substances encourage phagocytosis or bring about destruction of microbial membranes.

Two general pathways for complement activity exist. The **classical pathway** operates with the highly specific immune system and is initiated when certain antibodies unite with antigens and stimulate the complement system into action. The **alternative pathway** is nonspecific and is initiated by tumors, cell wall components of bacteria, and various microorganisms. It is sometimes called the **properdin pathway** because properdin is one of the proteins operating in it. The alternative pathway invokes a slower and less specific method for ridding the body of parasites, particularly Gram-negative bacteria and viruses.

Immunity is a state of specific resistance to infection. Specific resistance is directed against a particular type of microorganism and is the single most important characteristic of immunity.

The **immune system** enables the body to recognize a foreign agent as **nonself,** which is something other than a person's own substances **(self).** The immune system takes a specific action for neutralizing, killing, and eliminating that agent. The action involves nonspecific resistance as well. On occasion, the immune system activity may lead to tissue damage as seen in allergic disorders and other states of hypersensitivity.

The immune system's activity is based on its ability to distinguish characteristic proteins or protein-linked components associated with alien substances. Once this distinction has been made, certain lymphocytes are provoked to produce antibodies directed against the foreign matter, while other lymphocytes are sensitized to the invading agent and react with it directly. Thus, there are two major branches of the immune system: antibody-mediated immunity (also known as humoral immunity) and cell-mediated immunity.

Antigens

Immune responses are directed at a series of foreign substances known as **antigens,** also referred to as **immunogens.** Most antigens are high molecular weight substances, but low molecular weight substances will also act as antigens if they bind to proteins in the body. The low molecular weight compound is referred to as a **hapten.** The resulting conjugate may induce an immune response directed against the antigen.

The uptake and processing of antigens by macrophages in the tissue is an initial, critical step in most immune responses. The simple act of taking foreign substances into the body does not necessarily invoke an immune response because the substances may be broken down before they are ingested by macrophages.

Antigenic determinants. The chemical groups on the antigen molecules that determine their immunogenicity are called **antigenic determinants,** also known as **epitopes.** Antigenic determinants may consist of several amino acids of a protein molecule or several monosaccharide units of a polysaccharide. Each species of living thing is chemically and antigenically unique because of differences in its proteins, carbohydrates, and other organic substances.

Types of antigens. Certain types of antigens are distinctive. **Autoantigens,** for example, are a person's own self antigens. **Alloantigens** are antigens found in different members of the same species (the red blood cell antigens A and B are examples). **Heterophile antigens** are identical antigens found in the cells of different species.

A single organism such as a bacterium may contain a variety of proteins, carbohydrates, and other materials that provoke immune responses. Antigens found on the body cell are called **somatic antigens.** Antigens in the bacterial capsule are **capsular antigens.** Antigens of an organism's flagella are known as **flagellar antigens** (H antigens). Protein substances such as **exotoxins** are also antigenic.

Cells of the Immune System

Cells of the immune system are associated with the lymphatic system of the body and its specialized cells. Lymphocytes of the lymphatic system are derived from **stem cells** of the bone marrow. These undifferentiated precursor cells proliferate throughout life and replenish the mature cells of the immune system.

B-lymphocytes and T-lymphocytes. There are two major pathways for the differentiation of stem cells into immune cells. Certain of the stem cells produce **B-lymphocytes** (B-cells) while other stem cells form **T-lymphocytes** (T-cells). B-lymphocytes are so named because in birds, they are formed in the bursa of Fabricius. The equivalent site in humans has not been identified but is believed to be the bone marrow. T-lymphocytes undergo their conversion in the thymus gland, an organ in the neck tissues near the trachea and thyroid gland (Figure 30).

The transformation of stem cells into B-lymphocytes and T-lymphocytes begins about the fifth month after fertilization, and a full set is complete a few months after birth. These cells then migrate to the lymphoid organs in the lymph nodes, spleen, tonsils, adenoids, and other organs of the lymphatic system.

To initiate the immune response, microorganisms are phagocytized and their antigens are processed in phagocytic cells such as macrophages. The antigenic determinants are displayed on the surface of the phagocytic cells and presented to the appropriate B-lymphocytes and T-lymphocytes to provoke an immune response.

Clonal selection. The **clonal selection theory** helps explain how lymphocytes recognize antigenic determinants and respond. According to this theory, small populations (clones) of lymphocytes bear receptors on their cell membranes. Production of these receptors is genetically determined. On B-lymphocytes, the receptors consist of antibody molecules, while on T-lymphocytes, they are clusters of amino acids. When lymphocytes encounter an antigenic determinant on the surface of a macrophage, their receptors match with the antigenic determinant and a stimulation follows. A match is also made between a set of molecules called **major histocompatibility (MHC) molecules** and their receptors.

The clonal selection theory suggests that B-lymphocytes and T-lymphocytes exist for all antigenic determinants even before contact with an antigen is made. The theory also says that antigenic determinants stimulate the lymphocytes to endow their progeny with identical specificity. The B-lymphocytes and T-lymphocytes that might

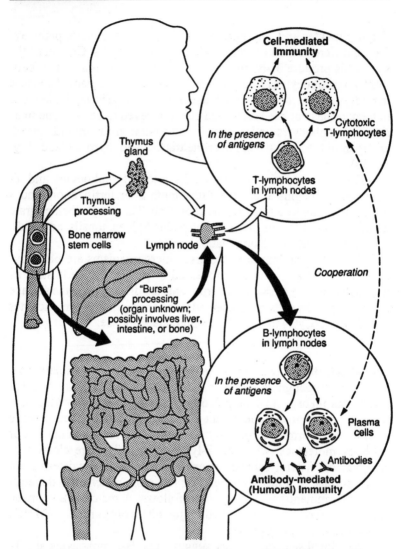

An overview of the human immune system. T-lymphocytes and B-lymphocytes originate in the bone marrow and then are processed in different body organs before proceeding to the lymph nodes, where they provide the underpinnings of the two types of immune responses.

■ Figure 30 ■

potentially react with the body's own cells are deleted or in some way inactivated to ensure that an immune response to the host organism does not develop.

Two general types of immunity exist for specific resistance to disease. They are antibody-mediated (humoral) immunity, centered in B-lymphocytes, and cell-mediated immunity, centered in T-lymphocytes.

Antibody-Mediated (Humoral) Immunity (AMI)

On exposure to antigenic determinants in lymphatic organs, B-lymphocytes are activated and differentiated to form **plasma cells.** Plasma cells are specialized, differentiated cells that synthesize and secrete antibodies specific for an antigen. Other activated B-lymphocytes form **memory cells.** These cells can be activated later to differentiate to plasma cells for rapid antibody production. This antibody production will occur on future reentry of the antigen to the body and is the basis of **long-term immunity.**

The products of plasma cells are **antibodies.** An antibody is a specialized protein substance produced by the host cells in response to an antigen in the host's tissues. Antibodies are capable of reacting specifically with the antigen that provoked their production. Antibodies are often referred to as **immunoglobulins.** They circulate in the blood and are associated with the gamma globulin fraction of the plasma.

Structure of antibodies.   Structurally, the antibodies (immunoglobulins) are composed of four protein chains arranged in a distinctive pattern. Each molecule has two long chains of heavy molecular weight protein (**H chains**) and two short chains of light molecular weight protein (**L chains**). The chains are linked by sulfur bonds.

At the outer end of each arm of the antibody molecule, a specific amino acid sequence exists. This is where the antibody molecule reacts with the antigenic determinant that provoked its production. The combining site is known as the **Fab region.** The most common antibody molecules have two Fab regions and are said to be **bivalent** (having two combining sites). The remaining portion of the antibody molecule is called the **Fc region** because it can be crystallized. Its amino acid content is relatively constant and characteristic for its class. This portion of the molecule activates the complement system and encourages phagocytosis.

When the antibody molecule reacts with the antigen, the two surfaces fit together like the pieces of a jigsaw puzzle. This "recognition" is exquisitely accurate and accounts for the extreme specificity of antibody molecules.

Classes of antibodies. Five classes of antibodies (immunoglobulins) are produced by the plasma cells. The first class, called **IgM,** is the major component of the primary antibody response in adult humans and is the first antibody to appear in the immune reaction. It is composed of five units joined by a J (joining) chain (Figure 31). IgM cannot diffuse through cell membranes and is found almost exclusively in the blood. Because of its many binding sites, it is more reactive with antigens than are other kinds of antibodies. IgM is also formed by the fetus during uterine development and is active against the A and B factors of the red blood cells. Many antitoxins formed against bacterial toxins are composed of IgM.

The principal antibody of the secondary immune response is **IgG.** This antibody is the most common in the bloodstream and is found in many secretions, such as spinal, synovial, lymph, and peritoneal fluids. IgG crosses the placenta and protects the fetus and newborn. IgG therefore provides a natural type of passive immunity. IgG also forms in the primary antibody reaction after a large amount of IgM has already formed. IgG has the "typical" antibody structure of four protein chains.

The third class of antibody is **IgA.** This antibody is found in external secretions such as those at the mucosal surfaces of the respiratory, gastrointestinal, and urogenital tracts. It is also present in the tears, saliva, bile, urine, and colostrum, and it is transferred in the

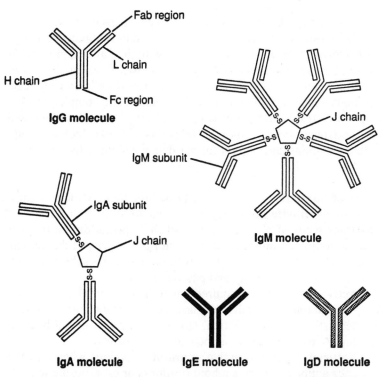

Details of an antibody molecule and the structures of the five types of antibody molecules produced by the human body.

■ Figure 31 ■

breast milk. IgA consists of two antibody units held together by a J chain and secretory component. IgA resists infections at the body surface.

The fourth class of antibody, **IgD,** is found in extremely small quantities in the serum. IgD is found at the surface of B-lymphocytes and is believed to be an antigen receptor at this location. IgM is also believed to be a receptor.

The final kind of antibody, **IgE,** occurs in minute concentrations in the serum and is important in hypersensitivity reactions, especially the anaphylactic reaction. A localized reaction is called allergy.

The reaction of antibodies with antigens helps neutralize the antigen and restrict the spread of infection. Certain antibodies react with the bacterial surface, while others react with the flagella, pili, or capsules. These reactions encourage phagocytosis. When antibodies react with a virus, viruses cannot attach to host cells and replicate. Antibody reaction with antigens also sets off the complement system, which results in the formation of an attack complex along with increased phagocytosis. Antibody reactions with toxin molecules neutralize the toxins and prevent further damage to body tissues.

Types of immunity. Immunity may be broadly classified as innate or acquired. **Innate immunity** is present from birth. It consists of numerous types of nonspecific factors that operate during times of disease. **Acquired immunity** is derived from activity of the immune system. The term generally refers to antibodies and is subdivided into two parts: active immunity and passive immunity.

Active immunity is acquired when the body produces antibodies. The immunity is usually long lasting because the immune system has been stimulated into action. However, it takes several hours to develop. Active immunity can be natural or artificial.

Naturally acquired active immunity develops when a person produces antibodies during a bout of illness or on exposure to a microorganism even though disease does not occur. The B-lymphocytes and plasma cells function, and this immunity occurs during the "natural" scheme of events.

Active immunity can also occur by artificial means. **Artificially acquired active immunity** occurs when a person produces antibodies after exposure to a vaccine. A **vaccine** consists of bacteria, viruses, or fragments of these. A vaccine may also contain **toxoids,** which are chemically treated bacterial toxins. Toxoid vaccines are available against diphtheria and tetanus. Viral vaccines are available against measles, mumps, rubella, polio, rabies, hepatitis A, hepatitis B, and yellow fever. Because vaccine exposures do not happen in the natural scheme of events, the immunity is said to be artificial.

Passive immunity comes about when the body receives antibodies from an outside source. In passive immunity, the immune system does not operate and the immunity is not long lasting. However, it protects instantaneously. Passive immunity can be either natural or artificial.

Naturally aquired passive immunity develops when antibodies pass from mother to child across the placental lining. The newborn is born with IgG, and it receives maternal IgA if it is breast-fed. These antibodies will remain for approximately six months after birth before fading away.

For **artificially acquired passive immunity,** a person is given an injection of antibodies. Usually these antibodies are derived from the blood of another individual who has produced antibodies when confronted with a certain disease. The antibody injected is usually IgG. The antibodies will remain for a period of several days or weeks and then disappear. This type of immunity is given to protect people who have been exposed to tetanus, diphtheria, or botulism. It is also used to protect against certain serious viral diseases. Allergic reactions to the serum proteins (serum sickness) limit the use of the preparations.

Cell-Mediated Immunity (CMI)

Cell-mediated immunity depends upon the activity of T-lymphocytes. T-lymphocytes have a longer life span than B-lymphocytes and are found in the same lymphatic tissues as the B-lymphocytes. The T-lymphocytes react with certain antigenic determinants and become immunologically "committed." Part of this commitment is the conversion to a subset of cells called cytotoxic T-lymphocytes.

Activity of cytotoxic T-lymphocytes. Cytotoxic T-lymphocytes do not produce antibody molecules. Rather, they leave the lymphatic tissues and enter the circulation. They circulate through the blood vessels and gather at the infection site. Here they interact directly with organisms such as fungi, protozoa, cancer cells, and transplant cells. They also interact with virus-infected cells and bacteria-infected cells (such as lung cells infected with tuberculosis). The T-lymphocytes exert a "lethal hit" on the cells and secrete substances into them that lead to cellular destruction.

In addition to their direct interaction, T-lymphocytes also secrete substances called **lymphokines.** Lymphokines attract phagocytes to the area and encourage them to perform phagocytosis on fungi, protozoa, and infected cells. This activity helps relieve the infection. Lymphokines are also known as **cytokines.** An important cytokine is **interleukin-1,** which activates T-lymphocytes, causing them to proliferate further and form clones.

Helper and suppressor T-lymphocytes. Helper T-lymphocytes also function in the immune system by encouraging the activity of B-lymphocytes in the production of antibodies. **Suppressor T-lymphocytes** regulate or suppress the activity of the immune system so that it is not excessive. **Natural killer (NK) cells** are T-lymphocytes that recognize and destroy many types of target cells without being exposed to antigens. Technically, these are not part of the specific immune response. Finally, the **delayed hypersensitivity T-lymphocytes** function in hypersensitivity reactions and encourage local tissue inflammations.

Laboratory Tests

Certain laboratory tests are available to detect the presence of antibodies in an individual. These laboratory tests are commonly used in diagnostic procedures because the presence of a certain antibody indicates the presence of a certain disease. The study of serum for its antibody content is known as **serology.**

One of the objects of serology is to determine the titer of antibody present in the individual. The **titer** is an estimate of the antibody level in a unit volume of serum. It is determined as a reactive dilution of the serum and is usually expressed in a ratio such as 1:100.

The agglutination and precipitation tests. Serological tests that involve clumping of an antigen are called **agglutination tests.** Antibodies involved in agglutination tests are called **agglutinins.** When the antibody molecules unite with antigen molecules on the surface of bacteria, red blood cells, or particles, they cause the cells to stick together and form large clumps (Figure 32).

Another form of serological test is the **precipitation test.** In this test, antibodies are called **precipitins.** They react with dissolved antigens and form large complexes that become visible as a fine precipitate. Tests such as these can be performed in fluid or gel.

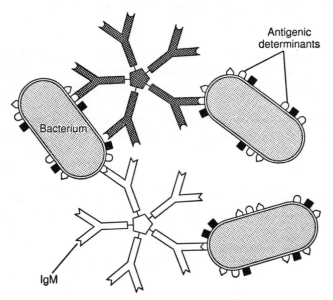

The agglutination reaction taking place when antibody molecules unite with the antigen molecules on the surfaces of cells and involve the entire cells in the reaction.

■ Figure 32 ■

The complement fixation test. The **complement fixation test** is another application of a serological reaction. In this test, antibodies are combined with antigens in the presence of complement, a series of blood proteins. If the antibodies are specific for that antigen, they will react and use up ("fix") the complement. When a subsequent combination of blood cells and their antibodies are added to the tube, no complement will be available and no reaction will take place. However, if the first antibodies were not specific for their antigen, the complement is not fixed, and it is available to react with the blood cells and their antibodies. The blood cells will undergo lysis as a result of the complement and antibody activity.

The neutralization test. Serological tests that determine the presence of antitoxins are called **neutralization reactions.** Antitoxins are antibodies formed against toxins. When these antibodies are united with toxin molecules, they neutralize the toxins and will prevent their toxic effects in an animal. However, if the antitoxin is not specific for the toxin, the toxin is not neutralized and it will exert a lethal effect on the animal.

The fluorescent antibody test. The **fluorescent antibody technique** is a serological test that helps to make visible an antigen-antibody reaction. Antibody solution is combined with cells that contain antigens. If an antibody-antigen reaction occurs, a fluorescent dye accumulates on the surface, thereby signaling a positive reaction.

Enzyme immunoassays. In the **enzyme immunoassay,** an antigen-antibody reaction is encouraged on the surface of a particle, and an enzyme accumulates there. When a substrate is added, the enzyme molecules change the color of the substrate, and the reaction can be designated positive. This test is sometimes called the **enzyme-linked immunosorbentassay (ELISA).**

A contemporary development in immunology is the development of **monoclonal antibodies.** Monoclonal antibodies are produced by a single colony of immortal cells, that is, cells with the ability to live and multiply for extremely long periods. The cells producing the antibodies are hybrids formed by fusing antibody-producing plasma cells with tumor cells (the immortal cells). Monoclonal antibodies can be purified and used for diagnostic purposes as well as carriers for toxic chemical compounds . The process can possibly be used to kill tumor cells in the body if antitumor drugs are carried.

Certain human disorders are attributed to activity of the immune system. These disorders are commonly known as hypersensitivities, states of increased immune sensitivity that are mediated by antibody or cellular factors. The disorders may also involve immunodeficiencies in which failures of antibody-mediated or cell-mediated immunity take place.

Hypersensitivity Reactions

Normally the immune system plays an important role in protecting the body from microorganisms and other foreign substances. If the activity of the immune system is excessive or overreactive, a **hypersensitivity reaction** develops. The consequences of a hypersensitivity reaction may be injury to the body or death.

Most injury resulting from hypersensitivities develops after an interaction has taken place between antigens and antibodies or between antigens and sensitized T-lymphocytes. The general nature of and symptoms accompanying the reaction depend upon whether antibodies or sensitized T-lymphocytes are involved. When antibodies are involved, the reactions fall under the heading of **immediate hypersensitivity.** When T-lymphocytes are involved, the reactions are characterized as **delayed hypersensitivity.** Immediate hypersensitivity reactions include anaphylaxis, allergic reactions, cytotoxic reactions, and immune complex reactions. Delayed hypersensitivity reactions are generally characterized as contact dermatitis or infection allergies.

Immediate hypersensitivity. The reactions accompanying immediate hypersensitivity depend upon the nature of the antigen, the frequency and route of antigen contact, and the type of antibody reacting with the antigen. The initial dose of antigen is referred to as the **sen-**

sitizing dose. This exposure is followed by a **latent period** and then a later dose of the same antigen, called the **eliciting dose** or **shocking dose.** The shocking dose sets off the hypersensitivity reaction, resulting in tissue damage.

Immediate reactions begin within minutes of contact with the eliciting dose of antigen. If antigens are introduced directly into the tissues, such as by insect sting or injection, the result is a systemic reaction such as **anaphylactic shock.** When the contact is a superficial one involving the epithelial tissues, the reaction is more localized, as occurs in asthma or allergic rhinitis (hay fever). These local reactions are commonly referred to as **allergy.** Another term used is **atopy.**

The antigens eliciting an immediate hypersensitivity are called **allergens,** particularly when they are involved in local allergic reactions. Hapten molecules such as penicillin molecules may be involved when they are bound to larger protein molecules. Foods, feathers, pollen grains, animal dander, and dust may be allergens. Animal sera, bee venoms, and wasp venoms are also allergens.

The antibodies involved in anaphylaxis reactions are of the type IgE. In cytotoxic and immune complex reactions, IgG and IgM are involved.

Anaphylaxis. Anaphylaxis, or **type I hypersensitivity,** is a whole-body, immediate hypersensitivity also known as **anaphylactic shock.** The allergens are introduced to the body directly to the tissues in a concentrated form (intramuscular or intravenous injection, for example).

After the sensitizing dose has been administered, IgE is produced by the plasma cells. The antibodies circulate in the blood and attach at the Fc end to **mast cells** of the tissues and **basophils** in the bloodstream (Figure 33). This activity occurs during the latent period. When the eliciting dose of allergen is later administered, the antigens combine with antibodies on the surface of the mast cells and basophils.

After the antigen-antibody combination has taken place, the cells release a number of physiologically active substances including **histamine** and **serotonin.** These substances are derived from granules

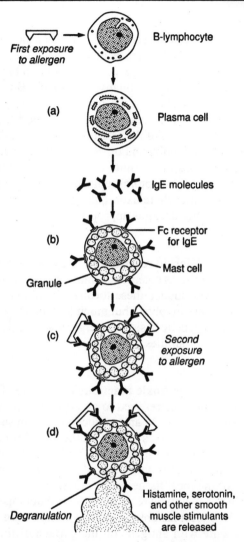

*The process of anaphylaxis. (a) Allergens stimulate the
production IgE antibodies, which (b) fix themselves to the
surfaces of mast cells. (c) On second exposure to the allergens,
a reaction occurs on the mast cell surface, and (d) the cellular
granules release histamine and other stimulators of smooth
muscle contraction.*

■ Figure 33 ■

within the cell. The histamine, serotonin, and other mediators induce spasms of the smooth muscle, such as in the bronchioles, small arteries, and gastrointestinal tract lining. A sudden drop in blood pressure occurs, followed by circulatory collapse and shock. Bronchospasms and edema cause constriction of the respiratory passageways, and breathing is very difficult. Facial edema occurs, and the heart rate increases due to constriction of the arteries. Swellings called "hives" develop at the site of injection and other areas of the skin. In severe cases, anaphylactic shock may result in death within several minutes to an hour. To relieve the symptoms, epinephrine is administered together with a smooth muscle relaxer, a drug such as cortisone to reduce swelling, and other drugs as appropriate.

Allergic reactions. **Allergic reactions** (allergy) are a milder, localized form of anaphylaxis. As noted, such things as foods, pollen grains, and animal dander can induce these localized reactions. IgE, basophils, and mast cells are involved, but much less than in anaphylaxis. There appears to be a genetic basis for allergic reactions, as evidenced by their distribution in families.

Cytotoxic reactions. **Cytotoxic reactions** are a form of immediate hypersensitivity, sometimes referred to as **type II hypersensitivity.** In these reactions, IgE and IgM are produced in response to stimulation by antigens. The antibodies unite with the antigens in the bloodstream, but they also unite with analogous antigens on the surface of the human body's cells. This union sets off the complement system, and destruction of the local tissue cells ensues.

An example of a cytotoxic reaction is **thrombocytopenia.** In this disease, antibody molecules are elicited by certain drug molecules. The antibodies unite with antigens on the surface of thrombocytes (platelets), and with complement activation, the thrombocytes are destroyed. The result is an impaired blood-clotting mechanism.

Another example of the cytotoxic reaction is **agranulocytosis.** In this immune disorder, antibodies unite with antigens on the surface of neutrophils. As these cells are destroyed with complement activation, the capacity for phagocytosis is reduced.

Cytotoxic reactions are also manifested by the **transfusion reaction** occurring when improper blood transfusions are performed. Another consequence is **erythroblastosis fetalis,** also known as **hemolytic disease of the newborn,** or Rh disease. In this condition, a pregnant woman produces Rh antibodies against the developing fetus, and when the Rh antibodies unite with Rh antigens on the surface of fetal red blood cells in a succeeding pregnancy, the red blood cells are destroyed (Figure 34).

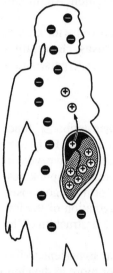 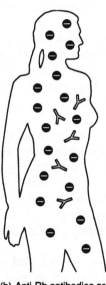 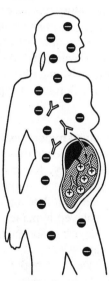

(a) Rh-positive erythrocytes from fetus enter blood of Rh-negative woman during the birth process.

(b) Anti-Rh antibodies are produced by woman's immune system that remain in the woman's bloodstream.

(c) During succeeding pregnancy, antibodies pass placental barrier and enter fetal blood causing the destruction of fetal erythrocytes.

The cytotoxic reaction in erythroblastosis fetalis.

■ Figure 34 ■

Immune complex disease. Immune complexes are combinations of antigen and antibody that have the ability to fix complement. The antibodies involved are IgM or IgG, and the antigens exist in fluid as soluble antigens. Proteins or nucleic acids may be involved.

An example of immune complex hypersensitivity is **serum sickness.** In this condition, animal serum is administered to humans, and its proteins elicit antibody production. When the antibodies and antigens unite, they form immune complexes, which activate the complement system and cause local tissue damage. The patient may display edema of the hands, face, and feet, as well as swelling of the upper respiratory tissues and impairment of normal respiration. An inflammatory response results.

Formation of immune complexes is also involved with numerous diseases including **systemic lupus erythematosus, rheumatoid arthritis,** and **glomerulonephritis.** Immune complex hypersensitivity is often called **type III hypersensitivity.**

Delayed hypersensitivity. T-lymphocytes rather than antibodies function in cases of **delayed hypersensitivity,** also called **type IV hypersensitivity.** Normally these are the T-lymphocytes involved in cell-mediated immunity. The T-lymphocytes produce lymphokines, which stimulate an influx of macrophages to perform phagocytosis. In delayed hypersensitivity, the result is an exaggeration of the immune response, and the phagocytes bring about the destruction of the local tissue.

Delayed hypersensitivity (also called cellular hypersensitivity) is so named because the reaction requires a day or more to develop. One manifestation of the reaction is **infection allergy,** as in the tuberculin skin test. A purified protein derivative (PPD) of *Mycobacterium tuberculosis* is applied to the skin superficially, and a skin reaction (swelling and redness) occurs 24 to 48 hours later if the person has had a previous exposure to the antigens of *Mycobacterium tuberculosis,* possibly during an episode of tuberculosis.

A second manifestation of delayed hypersensitivity is **contact dermatitis.** In many cases, the reaction is accompanied by large, blisterlike lesions in which vesicles are surrounded by a zone of erythema (redness). Usually, the vesicles itch intensely.

Antigens involved in contact dermatitis include metals such as nickel and mercury, cosmetics, disinfectants, and plant substances such as the resins of poison ivy, poison oak, and poison sumac. The individual can be tested to determine which antigen is the cause of allergy by performing a **patch test.** In this procedure, a patch containing a specific antigen is attached to the skin and left in place for 48 hours to determine if a reaction will take place.

Immunodeficiency Diseases

Disorders of the immune system may also be due to deficiencies of the system. These deficiencies may be **congenital** or **iatrogenic** (induced by immunosuppressive drugs), or they may result from malignancies occurring in the lymphatic system.

Abnormalities of the B-lymphocytes result in an immunodeficiency accompanied by abnormal production of antibodies. These abnormalities may result from infections such as diseases of the liver, from immune complex disorders such as systemic lupus erythematosus, or from malignancies. Multiple myeloma is a malignant disease of plasma cells in which certain clones of B-lymphocytes produce an overabundance of antibodies. The antibodies are excreted into the urine, where they are known as Bence Jones proteins.

Decreased ability of the B-lymphocytes to produce antibodies is called **hypogammaglobulinemia.** Such a condition may result from a genetic defect or the failure of the body to produce B-lymphocytes. Bone marrow disorders may also be a cause.

Abnormalities of the T-lymphocytes result in reduced capacity for cell-mediated immunity. This condition, called **DiGeorge's syndrome,** often develops from malformation or absence of the thymus gland. An individual with this syndrome is highly susceptible to infection by fungi, protozoa, and viruses.

When both B-lymphocytes and T-lymphocytes are deficient, the condition is called **severe combined immunodeficiency.** Recent procedures employing transplantation of bone marrow tissue have attempted to relieve this condition.

Microbial diseases of the skin are usually transmitted by contact with an infected individual. Although the skin normally provides a barrier to infection, when it is penetrated by microorganisms, infection develops. Diseases of the eye are considered with the skin diseases because both occur at the surface of the body.

Bacterial Skin Diseases

Staphylococcal infections. Staphylococci are Gram-positive cocci occurring in clusters. The best known pathogen in this group is *Staphylococcus aureus*. This organism invades the hair follicles and causes **folliculitis**, also referred to as **pustules**. A deeper infection of the skin tissues is referred to as a **boil, abscess,** or **furuncle**. These lesions are usually filled with pus. A large lesion progressing from a boil is known as a **carbuncle**. Infections such as these are easily transmitted by skin contact as well as by fomites.

Toxin-producing strains of *S. aureus* cause **scalded skin syndrome.** Usually found in young children and babies, this disease is characterized by vesicles on the body surface, which cause the skin to peel and give a scalded appearance. Penicillin or erythromycin antibiotics are used to treat this and other staphylococcal skin diseases.

Scarlet fever. **Scarlet fever** is caused by *Streptococcus pyogenes,* a Gram-positive bacterium occurring in encapsulated chains. Most cases of scarlet fever begin as infections of the respiratory tract, followed by spread of the bacteria to the blood. The bacteria produce an **erythrogenic toxin** that causes the typical skin rash. Penicillin is used for therapy. Complications include damage to the heart valves known as **rheumatic heart disease** or damage to the joints, which is called **rheumatic fever.**

Erysipelas. **Erysipelas** is a skin disease caused by *Streptococcus pyogenes* and other pathogenic streptococci. Small, bright, raised lesions develop at the site of streptococcal entry to the skin and grow with sharply defined borders. Penicillin therapy is employed.

Impetigo contagiosum. **Impetigo contagiosum** is a contagious skin infection accompanied by pus. It is caused by species of streptococci, staphylococci, and others. The disease commonly occurs in children and is easily transmitted among them. Penicillin therapy is often recommended.

Madura foot. **Madura foot** is a general name for infections of the feet due to many microorganisms. Among the causes are species of soil bacteria belonging to the genera *Nocardia, Actinomyces,* and *Streptomyces.* These and other bacteria enter the tissues and cause granular lesions that spread and eventually invade the bone and muscle. **Sulfurlike granules** represent accumulations of microorganisms in the pus, and antibiotic therapy is necessary to prevent spread of the disease.

Gas gangrene. **Gas gangrene** is a disease of the deep skin and wounds as well as the blood. Several species of *Clostridium* cause gas gangrene, including *Clostridium perfringens, C. novyi,* and *C. septicum.* These anaerobic rods are transferred to the wound in their spore form. They germinate and grow in the dead, anaerobic tissue of a wound, putrefying the proteins and fermenting the carbohydrates to produce gas. The gas causes the tissue to expand, and as the cells die from lack of oxygen, gangrene begins. Bacterial toxins pass through the bloodstream to cause illness throughout the body, and degeneration of the muscle fibers occurs. Aggressive antibiotic therapy and removal of dead tissue are useful therapies.

Cat scratch fever. **Cat scratch fever** may accompany a skin wound following a cat scratch. Although the causative agent has not been isolated with certainty, it is believed to be a species of *Rochalimaea* or *Afipia*. Patients display a pustule at the skin site of entry and swollen lymph nodes on one side of the body. Treatment with antibiotics may or may not be successful. Mild fever and conjunctivitis often accompany the disease.

Rat bite fever. **Rat bite fever** may be caused by either *Spirillum minor* or by *Streptobacillus moniliformis*. The former is a flagellated spiral bacterium; the latter is a Gram-negative rod in chains. Both species are transmitted during a bite by a rat, either wild or laboratory. Rat bite fever is associated with skin lesions, intermittent fever, and a skin rash. Arthritis may also be present.

Viral Skin Diseases

Rubella. **Rubella (German measles)** is a viral disease of numerous organs caused by an RNA virus and accompanied by a mild skin rash called an **exanthem.** First appearing on the body trunk, the rash spreads to other areas. Pregnant women may transmit the virus across the placenta to the developing embryo or fetus, and **congenital rubella syndrome** may develop in the newborn. Damage to the eyes, ears, and heart often result. Immunity can be rendered by an injection of attenuated rubella virus in the **MMR vaccine.**

Measles. **Measles** is also called **rubeola.** It is caused by an RNA virus normally transmitted by respiratory droplets during the coughing stage. Red spots with white centers occur on the cheeks, gums, and lips and are a diagnostic sign for the disease. These spots are called **Koplik spots.** The measles skin rash appears as a blush first on the forehead, then on the upper extremities, trunk, and lower extremities. Prevention is rendered by inoculation with attenuated measles viruses in the **MMR vaccine.** Complications of the disease may include measles encephalitis or subacute sclerosing panencephalitis (SSPE).

Chickenpox. Chickenpox is also called **varicella.** The disease is closely related to an adult disease called **herpes zoster (shingles).** The responsible virus is a DNA-containing virus of the herpesvirus group. It is also known as the **VZ virus.**

Chickenpox is a highly contagious disease. Transmitted primarily by respiratory droplets, the disease is accompanied by teardrop-shaped lesions filled with fluid. The lesions begin on the scalp and trunk and then spread to the face and limbs. Prevention is possible with injections of inactivated VZ virus in the chickenpox vaccine.

Shingles occurs in adults and is believed to be a recurrence of infection by the virus that causes chickenpox. Presumably, the virus has remained latent in ganglia of the nervous system until it is reactivated. The disease is characterized by painful lesions surrounding the body trunk. The disease is highly contagious. Acyclovir may be recommended for therapy.

Smallpox. Smallpox is a viral disease caused by a large, boxlike, DNA-containing virus having a complex shape. At one time, smallpox was a major cause of death in the world. It was accompanied by pus-filled lesions covering the body surface, and usually it resulted in death. Immunity was rendered by an injection of cowpox (vaccinia) viruses, as first recommended by Edward Jenner in 1798. Smallpox has apparently been eradicated on the earth and has not appeared in humans since October 26, 1977. It is the first infectious disease ever to be eradicated.

Cowpox. Cowpox, also known as **vaccinia,** is caused by a DNA virus similar in shape to the smallpox virus. In barnyard animals, the virus causes a disease accompanied by lesions of the skin. These lesions also occur when humans are infected. Immunizations with cowpox viruses for smallpox protection are no longer given.

Molluscum contagiosum. **Molluscum contagiosum** is a skin disease caused by a DNA-containing poxvirus. The disease is accompanied by flesh-colored, painless lesions scattered over the skin surface. The disease is transmitted by skin contact.

Warts. **Warts** are considered an infectious disease caused by a number of **papilloma viruses,** which contain DNA. Warts vary in appearance, and are generally benign. However, certain types of warts can be forerunners of malignancies. Cases of **genital warts** are very widespread, and certain strains of virus are related to cervical cancers. Genital warts are transmitted by sexual skin contact. Other kinds of warts, such as **dermal warts,** occur in the epithelial cells of the skin tissues.

Fungal and Parasitic Skin Diseases

Athlete's foot and ringworm. Both **athlete's foot** and **ringworm** are caused by various species of fungi belonging to the genera *Trichophyton, Microsporum,* and *Epidermophyton.* These fungi are often called **dermatophytes,** and their diseases are referred to as **dermatomycoses.** Both diseases are accompanied by fluid-filled lesions occurring on the body surface. The diseases are spread by fragments of fungal hyphae. Athlete's foot is also called **tinea pedis,** while ringworm may be called **tinea corporis** (ringworm of the body), **tinea cruris** (ringworm of the groin), or **tinea capitis** (ringworm of the scalp). Many pharmaceutical ointments are available to prevent spread of the disease, and the antibiotic griseofulvin is available by prescription.

Sporotrichosis. **Sporotrichosis** is caused by the fungus *Sporothrix schenckii.* The fungus is transmitted during skin wounds associated with thorns of rose or barberry bushes, as well as by contact with sphagnum moss. The disease is accompanied by a nodular mass at the site of entry; then it spreads to the lymphatic vessels and swelling (edema) follows. Hard, knotlike growths are found beneath the body surface. Potassium iodide and amphotericin B may be used for therapy.

Blastomycosis. Blastomycosis is a fungal disease due to *Blastomyces dermatitidis*. This fungus is transmitted from the lungs of an infected patient or from a wound. In a wound it causes pus-filled lesions and multiple abscesses. A systemic form of blastomycosis may develop, with involvement of other organs. Amphotericin B is used for severe cases.

Candidiasis (yeast disease). The fungus *Candida albicans* is commonly found in the normal flora of numerous body tracts, but in compromised individuals, it may cause a superficial infection known as **candidiasis** or **yeast disease.** Yeast disease occurs in the vaginal tract and is accompanied by internal discomfort, pruritis (itching sensations), and sometimes, a discharge. Yeast disease often follows the destruction of lactobacilli in the vaginal tract. It can be treated with such drugs as miconazole, ketoconazole, and itraconazole.

Candida albicans may also cause infection in other skin locations. For example, **thrush** is a form of candidiasis in which patches of inflammation occur on the tongue and mucous membranes of the mouth. A skin infection called **onchyosis** occurs in individuals whose hands are in contact with water for long periods.

Swimmer's itch. Swimmer's itch is a skin infection due to tissue invasion by species of the flatworm *Schistosoma.* The schistosomes are not pathogenic of themselves, but they induce an allergic reaction that brings on the skin irritation and itching associated with the disease. Transmission occurs during swimming in contaminated waters.

Dracunculiasis. Dracunculiasis is a skin disease caused by the round-worm *Dracunculus medinensis.* In this disease, the roundworms live in skin lesions and emerge through the lesions. In tropical countries, dracunculiasis is widespread, and relief from the disease consists of removing the roundworms through openings made in the lesions.

Eye Diseases

Conjunctivitis. Conjunctivitis is a general term for infection of the membrane covering the inner eyelid and pupil of the eye. This membrane is called the conjunctiva. **Bacterial conjunctivitis** is also known as **pinkeye.** It is caused by numerous bacteria, most commonly the Gram-negative rod *Haemophilus aegyptius.* The disease is characterized by red, itchy eyes with an exudate. It is highly contagious and is transmitted by droplets and contact to other individuals. Various ointments and fluids containing neomycin are used for therapy.

Trachoma. Trachoma is a bacterial infection of the eye caused by *Chlamydia trachomatis.* This organism is an extremely tiny chlamydia. It causes an infection of the cornea in which rough, sandy, pebblelike growths occur and interfere with vision. Tetracycline and other antibiotics are used for treatment. Transmission usually occurs by contact.

Secondary eye infections. Many sexually transmitted diseases result in secondary eye infections of the newborn when the bacteria are contacted during the birth process. One example of an infection is **gonococcal ophthalmia,** caused by *Neisseria gonorrheae,* the organism of gonorrhea. Inflammation of the cornea in the newborn can lead to blindness. Another possibility is **chlamydial ophthalmia,** due to infection with *Chlamydia trachomatis,* the organism that causes chlamydia. Antibiotics are used to treat these infections, and the eyes of newborns are routinely treated with antibiotic to prevent their occurrence.

Herpes keratitis. **Herpes keratitis** is caused by the herpes simplex virus, which has DNA. Transmitted by contact, this virus causes lesions of the cornea and other eye structure and may cause blindness. Acyclovir is used for therapy.

Adenoviral keratoconjunctivitis. **Adenoviral keratoconjunctivitis** is caused by a DNA virus called the **adenovirus.** This virus normally causes the common cold syndrome, but it can also be transmitted to the eye, where it may cause corneal opaqueness. When transmitted by water, the infection is called **shipyard eye.**

Loaiasis. **Loaiasis** is caused by the eyeworm *Loa loa.* This roundworm is transmitted among humans by deerflies. The worms live in the skin tissues and concentrate in the conjunctiva and cornea of the eye. They can be removed with optical instruments.

Microbial diseases affecting the nervous system tend to be serious because of the critical functions performed by the brain, spinal cord, and peripheral and cranial nerves. Infections can occur in the nervous tissue or in the covering membranes called meninges. Diagnostic tests for diseases of the nervous system often involve examination of the cerebrospinal fluid, and antibiotic therapy must use drugs that pass the blood-brain barrier.

Bacterial Diseases

Meningococcal meningitis. **Meningococcal meningitis** is caused by the Gram-negative diplococcus *Neisseria meningitidis.* This organism is called the **meningococcus.** It is transmitted by respiratory droplets and often inhabits the nasopharynx without evidence of disease. The organism is believed to possess endotoxins that account for the symptoms associated with meningitis. Patients suffer severe and debilitating headache, as well as fever, chills, and blue-black skin spots. The neck is stiff, and seizures are possible. Examination of the cerebrospinal fluid reveals Gram-negative diplococci. The adrenal glands may by involved (**Waterhouse-Friderichsen syndrome**). Aggressive therapy with penicillin and other drugs is required.

Haemophilus **meningitis.** *Haemophilus* **meningitis** is caused by *Haemophilus influenzae* **type b.** The organism is a Gram-negative, small rod that usually affects children during the first year or two of life. Nerve disorder, fever, and possible mental retardation result from the disease. The **Hib vaccine** is used to provide immunity, and rifampin and other antibiotics are used in therapy.

Listeriosis. Listeriosis is caused by a small, Gram-positive bacterium called *Listeria monocytogenes.* Also a blood disease, listeriosis can affect the meninges (**listeric meningitis**). The disease is transmitted by unpasteurized or improperly pasteurized milk and cheese products, as well as from animals. In a pregnant woman, the bacillus may affect the fetus and cause miscarriage.

Leprosy. Leprosy is considered a disease of the nervous system because the bacilli destroy the peripheral nerves in the skin. Thus affected, the patient cannot sense environmental changes, and injury to the skin tissues results. Deformed hands and feet and eroded bones, fingers, and toes are seen in the disease. In **tuberculoid leprosy**, skin pigments are lost. In **lepromatous leprosy**, skin nodules called **lepromas** disfigure the skin. The incubation time is roughly three to six years.

Leprosy is caused by an acid-fast bacillus called *Mycobacterium leprae.* The organism is cultivated with great difficulty in the laboratory. The disease is known by its preferred name **Hansen's disease**. Dapsone is used for therapy.

Tetanus. Tetanus is caused by the soilborne, anaerobic, Gram-positive rod *Clostridium tetani.* Spores of this organism enter a wound, where they germinate to vegetative cells. The organisms produce a powerful **exotoxin** that interferes with the removal of acetylcholine from the synapses in the nervous system. This inhibition results in spasms affecting the muscles and causing clenched jaws and fists, paralysis of the respiratory muscles, disturbance of heart function, and death. The disease is prevented with immunizations of tetanus toxoid in the **DPT vaccine**. Established cases are treated with tetanus antitoxin (antibodies) and large doses of antibiotic such as penicillin.

Botulism. Botulism is caused by the anaerobic, Gram-positive, sporeforming rod known as *Clostridium botulinum.* The organism's spores enter food in vacuum-sealed, anaerobic environments, and they germinate to reproducing cells, which produce powerful **exotoxins** ingested with the food. The toxin interferes with the release of acetylcholine in the synapse between nerve and muscle cells. Without acetylcholine, nerve impulses cannot be transmitted, and paralysis soon begins. Respiratory arrest leads to death. No vaccine is available, but treatment with large doses of botulism antitoxin may prevent death. Infant botulism and wound botulism are also possible.

Viral Diseases

Rabies. Rabies is a viral disease of the brain that has a mortality rate approaching 100 percent. The agent is an RNA virus of the family Rhabdoviridae. Transmitted from warm-blooded animals, the rabies virus affects the brain, causing neurological distress and paralysis in muscles. Paralysis of the swallowing muscles results in **hydrophobia,** the fear of water. **Immunization** with inactivated viruses may be rendered after the virus has been transmitted in a bite. Four or five inoculations in the shoulder muscle are sufficient to induce immunity and prevent the development of symptoms. Pre-exposure vaccination is also possible.

Encephalitis. Encephalitis is an inflammation of the brain tissue, usually due to any of a variety of RNA-containing viruses. Among the kinds of encephalitis are **eastern equine encephalitis (EEE),** **western equine encephalitis (WEE),** and **Venezuelan eastern equine encephalitis (VEEE).** All are transmitted from horses by **arthropods** such as mosquitoes. Other forms of encephalitis include **St. Louis encephalitis, California encephalitis,** and **La Crosse encephalitis.** Patients suffer fever and severe headache, and fatalities are common. Control consists of killing the arthropods that transmit the viruses.

Poliomyelitis. Poliomyelitis (or **polio**) is a nervous system disease caused by an RNA virus belonging to the Picornaviridae family. The virus is usually transmitted by contaminated food and water and causes intestinal distress. Viruses then reach the central nervous system and may cause meningitis or paralysis if they reach the spinal cord. Prevention is available with the **inactivated virus vaccine (Salk)** or the **attenuated virus vaccine (Sabin)**. Both vaccines contain the three known strains of polio virus. **Postpolio syndrome** may occur in patients who experienced polio many years before. Weakened muscles and local paralysis characterize this condition.

Other Diseases

Sleeping sickness (trypanosomiasis). **Sleeping sickness** is also known as **trypanosomiasis** because the etiologic agent is a protozoan belonging to the genus *Trypanosoma*. The species responsible for **African** sleeping sickness is *T. brucei*, which is transmitted by tsetse flies and infects the blood of patients. Headache, lassitude, tremors, and uncoordinated movements characterize infection of the nervous system. Blood smears reveal the trypanosomes, and drug therapy is available with pentamidine and suramin.

The **South American** form of sleeping sickness is also known as **Chagas' disease.** The etiologic agent is *T. cruzi*. This trypanosome is transmitted by triatomid bugs. The organisms affect the nervous system of patients as well as the heart tissue. Often they destroy the nerve ganglia of the heart and cause severe heart disease.

Slow-developing diseases. Other diseases of the nervous system are believed due to viruses that have not yet been isolated. An example is **kuru,** a slow-developing disease observed in South Pacific peoples. Kuru is called a "slow virus disease" because the symptoms, which include nervous tremors, take over a year to appear. Similar diseases are **Creutzfeldt-Jakob disease, scrapie** (in goats and sheep), **bovine spongiform encephalopathy** (the "mad cow disease"), and a number of other diseases possibly caused by **prions**. Prions are protein particles that do not appear to have nucleic acid associated with them.

The infectious diseases of the cardiovascular system infect the blood, blood vessels, and heart. In many cases, the infections remain in these areas, but in others, the infections are spread to secondary organs. The diseases of the lymphatic system affect the lymph, lymph vessels, lymph nodes, and lymphoid organs, such as the spleen, tonsils, and thymus.

Bacterial Diseases

Streptococcal septicemia. Septicemia is a general expression for microbial infection of the blood and blood vessels. In previous generations, this condition was known as **blood poisoning.** A common cause of streptococcal septicemia is the Gram-positive streptococcus named *Streptococcus pyogenes.* This beta-hemolytic streptococcus causes severe fever, malaise, and dropping blood pressure. Shock may accompany the infection, and antibiotic therapy with penicillin is used aggressively. Septicemia may also be caused by a number of Gram-negative rods that release endotoxins.

An important complication of streptococcal septicemia is **endocarditis,** an infection of the heart valves. This is usually an immune system problem caused by antigen-antibody reactions taking place at the heart valves. Heart valve replacement is sometimes required. The subacute form due to *Streptococcus pyogenes* is accompanied by fever, weakness, and heart murmur. The acute form is generally due to infection by *Staphylococcus aureus* and is accompanied by rapid destruction of the heart valves.

Rheumatic fever is an immune reaction taking place in the heart tissues and is usually stimulated by antigens derived from *Streptococcus pyogenes.* Inflammation of the heart tissues is often accompanied by inflammation and arthritis of the joints, a condition called **rheumatoid arthritis. A streptococcal sore throat** may precede this condition.

Tularemia. **Tularemia** is due to a Gram-negative rod called *Francisella tularensis.* The bacteria enter the body by contact, inhalation, ingestion of contaminated rabbit meat, and the bite of ticks and other arthropods. Patients experience a blood disorder accompanied by fever, malaise, and numerous nonspecific symptoms. Antibiotics such as gentamicin are used in therapy.

Plague. **Plague** is caused by the Gram-negative rod *Yersinia pestis.* This organism is similar to the agent of tularemia and is transmitted by its rodent reservoir, the **rat flea.** The organism enters the lymphatic system and causes swelling of the lymph nodes called **buboes.** This stage is called **bubonic plague.** When the bacteria enter the blood, the condition is referred to as **septicemic plague,** and when the bacteria enter the lungs, the disease is called **pneumonic plague.** Transmission by airborne droplets is possible at this time. Aggressive antibiotic therapy is necessary to prevent death. The bacteria display a safety-pin appearance due to the accumulation of dye at the poles of the cells. This characteristic is called **bipolar staining.**

Brucellosis. **Brucellosis** is also known as **undulant fever** because it is characterized by alternating periods of high fever and relief. The bacterial agents belong to the genus *Brucella.* They are small, Gram-negative rods and include *B. abortus, B. suis, B. melitensis,* and *B. canis.* In animals, these bacteria cause abortion of the young (**contagious abortion**) and sterility of the female. They are transmitted to humans by unpasteurized milk and contaminated meat. On entering the bloodstream, the bacteria cause fever, chills, and malaise. Prolonged treatment is required with tetracycline, and vaccines are available for immunizing herds of animals.

Anthrax. **Anthrax** is due to the Gram-positive, aerobic, sporeforming rod *Bacillus anthracis.* Spores from this organism are inhaled from the air, or they are acquired during contact with contaminated soil or animals such as sheep and cattle. In the bloodstream, *B. anthracis* causes severe hemorrhaging, and the spleen, kidneys, and other blood-rich organs become engorged with blood. In the lungs, anthrax is called **woolsorter's disease** and is accompanied by pneumonia. Aggressive antibiotic therapy is necessary to prevent death.

Relapsing fever. **Relapsing fever** is so named because of the recurrent periods of fever. The etiologic agent is *Borrelia recurrentis,* which is a spirochete. The organism is transmitted by **lice,** which are natural parasites of humans. It may also be transmitted among humans by **ticks.** Jaundice and rose-colored skin spots accompany the infection, which may be treated by antibiotics.

Lyme disease. **Lyme disease** is caused by *Borrelia burgdorferi.* This organism is a spirochete transmitted by **ticks** of the genus *Ixodes.* First observed in Lyme, Connecticut, Lyme disease is now found throughout the United States.

Among the first symptoms of Lyme disease is a **bull's-eye rash** occurring on the skin. The rash is called **erythema chronicum migrans.** It occurs at the site of the tick bite and has a red center and expands over a period of several days. After the rash fades and spirochetes enter the blood, fever and other symptoms appear. In addition, the heart is affected and irregular heartbeat may be observed. On occasion, there is paralysis of the face and meningitis. Some months later, patients display arthritis of the large joints such as hips, ankles, elbows, and knees.

Lyme disease may be treated with a number of antibiotics, including penicillin and tetracycline. A vaccine is currently available for dogs. Diagnosis of the disease depends upon the observance of symptoms and awareness of exposure to ticks.

Rocky Mountain spotted fever. **Rocky Mountain spotted fever** is caused by the rickettsia *Rickettsia rickettsii.* This submicroscopic bacterium is transmitted by **ticks** of the genus *Dermacentor.* The disease is characterized by a **maculopapular skin rash** (a "spotted rash") occurring on the appendages and then spreading to the trunk. The fever is very high, and headaches accompany the disease. Antibiotics such as tetracycline are effective for therapy.

Epidemic typhus. **Epidemic typhus** is caused by *Rickettsia prowazekii,* a rickettsia transmitted by the **body louse** of the genus *Pediculus.* The organism invades the bloodstream and causes a **maculopapular skin rash** beginning on the body trunk and spreading to the appendages. The fever is extremely high, and the death rate is significant. Tetracycline antibiotics are effective for therapy, and elimination of lice is essential to stem the spread of the epidemic.

Endemic typhus. **Endemic typhus** is also called **murine typhus** because it occurs in mice and other rodents. It is transmitted by the **rat flea** and is caused by *Rickettsia typhi,* a submicroscopic rickettsia. The symptoms are similar to those of epidemic typhus but are much milder, and the mortality rate is much lower.

Other rickettsial diseases. Several other rickettsiae are known to cause diseases in humans. One example is **rickettsialpox,** caused by *Rickettsia akari.* This organism is transmitted by **mites** and causes a skin rash that resembles chickenpox. Another disease is **tsutsugamushi,** also called **scrub typhus.** This disease is also transmitted by **mites.** It occurs in Pacific regions and is characterized by a fever and skin rash.

Another rickettsial disease is **trench fever,** caused by *Rochalimaea quintana.* This disease is transmitted by **lice** and was common during World War I, when it affected soldiers in the trenches. **Ehrlichiosis** is a rickettsial disease due to *Ehrlichia canis.* Patients suffer headache and fever, but there is no skin rash associated with the disease. A similar disease is **human granulocytic ehrlichiosis (HGE),** which is also caused by a species of *Ehrlichia. Ehrlichia* species are transmitted by **ticks.** The diseases can be treated with tetracycline and other antibiotics.

Viral Diseases

Yellow fever. Yellow fever is a viral disease of the bloodstream transmitted by the **mosquito** *Aedes aegypti*. The virus is an RNA-containing particle that is icosahedral. After injection by the mosquito, the virus spreads to the lymph nodes and blood, where it persists in the blood-rich organs such as the liver. Very high fever, nausea, and jaundice accompany the disease. The mortality rate is high. Two vaccines are available for preventing yellow fever.

Dengue fever. Dengue fever is transmitted by the *Aedes aegypti* **mosquito** and caused by an RNA virus. The viruses enter the bloodstream, where they cause fever and severe muscle, bone, and joint pains, leading to **breakbone fever.** Successive exposures to the virus may result in **dengue hemorrhagic fever,** in which extensive hemorrhaging occurs in the blood-rich organs.

Infectious mononucleosis. Infectious mononucleosis is caused by a herpesvirus believed to be the **Epstein-Barr virus.** This virus has DNA and an envelope and the ability to remain latent in the B-lymphocytes. Symptoms of infectious mononucleosis include sore throat, mild fever, enlarged spleen, and an elevation of infected B-lymphocytes known as **Downey cells.** The viruses are often transmitted by saliva. Treatment usually consists of extensive bed rest, and recurrences are possible.

The virus of infectious mononucleosis is related to a type of tumor of the jaw tissues known as **Burkitt's lymphoma.** Most often seen in Africa, the condition is related to mononucleosis because of its etiologic agent. The Epstein-Barr virus is also related to cases of **Epstein-Barr virus disease,** known on occasion as **chronic fatigue syndrome.**

Acquired immune deficiency syndrome (AIDS). The **AIDS** epidemic was first recognized in the United States in 1981, when physicians in Los Angeles and other cities noted an unusually large number of opportunistic microbial infections. Destruction of T-lymphocytes of the immune system was associated with these infections. By 1984, the responsible virus had been identified, and in 1986, it was given the name **human immunodeficiency virus (HIV).**

HIV is a very fragile virus, and for this reason, it does not survive long periods of exposure outside the body. Most cases are transmitted directly from person to person via transfer of blood or semen. The disease is associated with intravenous drug users who use contaminated needles and with individuals who perform anal intercourse, since bleeding is often associated with this practice. Heterosexual intercourse can also be a mode of transmission, especially if lesions occur on the reproductive organs.

In the infected individual, HIV infects T-lymphocytes by combining its spike glycoproteins with the **CD4 receptor sites** of T-lymphocytes. The nucleocapsid enters the cytoplasm of the T-lymphocyte, and the viral enzyme **reverse transcriptase** synthesizes DNA molecules using the RNA of HIV as a template (for this reason, the virus is called a **retrovirus**).

The DNA molecule, known as a **provirus,** assumes a relationship with the DNA of the T-lymphocyte and enters the state of **lysogeny.** From this point, the provirus encodes new HIV particles. The human body attempts to keep up with the mass of new viral particles, but eventually, the newly emerging strains of HIV overwhelm the body defenses and the T-lymphocyte count begins to drop. Normally, the count is approximately 800 T-lymphocytes per cubic millimeter of blood, but as the disease progresses, it drops into the low hundreds and tens. This drop may occur as soon as six months after infection or as long as 12 years or longer after infection.

While the T-lymphocytes are infected, and so long as the T-lymphocyte level remains close to normal, the patient is said to have **HIV infection.** The patient occasionally will suffer swollen lymph nodes, mild prolonged fever, diarrhea, malaise, or other nonspecific symptoms. **AIDS** is the end stage of the disease. It is signaled by the appearance of **opportunistic infections** such as candidiasis, an excessively low T-lymphocyte count, a wasting syndrome, or deterioration of the mental faculties.

When a person has progressed to AIDS, an opportunistic infection is usually present. This infection may be *Pneumocystis carinii* pneumonia; *Cryptosporidium* diarrhea; encephalitis due to *Toxoplasma gondii;* severe eye infection and blindness due to cytomegalovirus; candidiasis of the mucous membranes and esophagus due to *Candida albicans;* meningitis due to *Cryptococcus neoformans;* or herpes simplex, tuberculosis, or cancer of the skin known as Kaposi's sarcoma. These opportunistic infections are treatable with various drugs, but the AIDS patient is constantly fighting one or the other, and it is difficult to retain the will to continue resisting. As of 1996, close to 600,000 cases of AIDS had been recognized in the United States, and approximately 400,000 patients had died.

Also as of 1996, two types of drugs were available to inhibit the multiplication of HIV. One group is the **chain terminators,** such as **azidothymidine (AZT), dideoxycytidine (ddC),** and **dideoxyinosine (ddI).** These drugs interfere with the synthesis of the DNA molecule using the viral RNA as a template. They effectively interfere with the activity of the reverse transcriptase. The second group consists of **protease inhibitors.** These drugs include saquinivir and indivir. They prevent the synthesis of the viral capsid by interfering with the last steps in preparation of the protein.

Diagnostic tests for AIDS are usually **antibody-based tests.** These tests seek to determine the presence of antibodies produced by the body on entry of HIV. It takes approximately six weeks for the body to produce sufficient antibodies for a positive test. Other tests called **antigen-based tests** are designed to detect the virus itself. These tests use gene probes that unite with and signal the presence of the viral DNA if it is present in the T-lymphocytes. Counts of the T-lymphocytes are performed by a process called flow cytometry.

Thus far, **vaccines** are not available against HIV. There is question, for example, whether whole viruses or viral fragments are preferred for the vaccine. Two glycoproteins called **gp120** and **gp41** from the envelope spikes are being investigated as possible vaccines. Tests are hampered however, since animal models are not available for vaccine testing, and it is difficult to find volunteers, who would then be antibody-positive and could suffer discrimination as a result. Nevertheless, candidate vaccines have been prepared not only with gp120 and gp41, but also with simian immunodeficiency virus (SIV), which infects primates, and viruses mutated so as to have no envelopes. Many candidate vaccines are now in the testing stage, and it is hoped that one will soon be available for the general population.

Protozoal and Parasitic Diseases

Toxoplasmosis. **Toxoplasmosis** is a protozoal disease caused by the sporozoan *Toxoplasma gondii.* This protozoan is transmitted from domestic **house cats,** usually by contact with their urine or feces. In humans, the protozoa multiply in the bloodstream and undergo a complex reproductive cycle. Patients experience fever, with other constitutional abnormalities, but symptoms are generally mild. However, in a **pregnant woman,** the protozoa may pass to the unborn fetus and cause tissue destruction. Also, in AIDS patients, toxoplasmosis can result in seizures and then brain inflammation, and it may be a cause of death.

Malaria. **Malaria** is a blood disease due to many species of the genus *Plasmodium.* Plasmodia are a group of protozoa of the Sporozoa (Apicomplexa) group. The parasites are transmitted by **mosquitoes** belonging to the genus *Anopheles.* When they infect individuals, they invade the red blood cells in the **merozoite** form. Within the red blood cells, the protozoa undergo various stages of their life cycle, and eventually the red blood cells rupture to release large numbers of parasites. The toxic compounds released during the rupture cause the paroxysms of chills and fever that characterize malaria. Severe anemia results, and renewed infections take place in new red blood cells. Treatment is effective with drugs such as quinine, chloroquine, and primaquine. The mortality rate remains high, however, and malaria infects approximately 300 million people each year.

Schistosomiasis. Schistosomiasis is caused by a multicellular, parasitic flatworm known as a **fluke.** The responsible flukes include *Schistosoma mansoni* and other species. In water, these parasites live in **snails,** and they enter the body through the skin of an individual who walks or swims in the infected water. The parasites multiply and live within the bloodstream, where they interfere with the flow of blood and lymph and cause local tissue damage. Various chemotherapeutic drugs are available to treat the disease.

Microbial diseases of the respiratory system may occur in the upper or lower regions. The upper region consists of the nose, pharynx, and other structures such as the middle ear and sinuses. Although many defensive mechanisms exist in this part of the body, such as ciliated hairs and mucous membranes, infections are common because of the proximity to the external environment. The lower portion of the system consists of the respiratory tubes and alveoli of the lungs. Infection occurs here because of the excessive moisture and rich supply of nutrients.

Bacterial Diseases

Strep throat. The common **strep throat** is due to a group A beta-hemolytic streptococcus known as *Streptococcus pyogenes.* This Gram-positive organism is encapsulated and produces streptokinase, which breaks down fibrin clots and permits the organism to spread to other tissues. The disease is accompanied by enlarged lymph nodes, inflamed tissues, and pus found on the tonsils. Diagnosis can be performed by obtaining throat swabs and combining the bacteria present with specific antibodies coated to beads. If the beads clump together, then *S. pyogenes* is presumably present. Cases of strep throat are treated with penicillin antibiotics.

Scarlet fever. **Scarlet fever** is caused by *Streptococcus pyogenes,* the same organism that causes strep throat. In scarlet fever, the beta-hemolytic streptococci produce an **erythrogenic toxin,** which causes a skin rash. The fever is usually high, the throat tissues are inflamed, and the tongue exhibits a strawberrylike appearance ("strawberry tongue"). Penicillin antibiotics are normally used in therapy.

Diphtheria. **Diphtheria** is caused by a club-shaped, Gram-positive rod called *Corynebacterium diphtheriae.* The disease is characterized by sore throat, neck swelling, and blockage of the respiratory passageways with membranelike accumulations. These accumulations are due to the effects of a bacterial **exotoxin,** which destroys cells of the epithelial lining. Antibiotic therapy is augmented by administration of antitoxins to neutralize the toxins. Immunization is imparted in the **DPT vaccine,** in which diphtheria toxoid is employed.

Otitis media. **Otitis media** is infection of the middle ear accompanied by earache. Numerous bacteria may cause this problem including *Streptococcus pneumoniae, Haemophilus influenzae,* and *Staphylococcus aureus.* Antibiotic therapy is usually indicated.

Pertussis (whooping cough). **Pertussis** is caused by the Gram-negative bacterium *Bordetella pertussis.* Transmitted by airborne droplets, this organism multiplies in the trachea and bronchi and causes paroxysms of cough. A rapid inrush of air following a paroxysm results in the high-pitched **whooping** sound. Treatment is rendered with erythromycin and other antibiotics, and immunization may be performed with killed pertussis bacilli in the **DPT vaccine** or acellular bacterial fragments in the **DTaP vaccine.**

Tuberculosis. **Tuberculosis** is caused by *Mycobacterium tuberculosis,* an **acid-fast rod.** The bacteria have large amounts of mycolic acid in their cell walls, which permits them to retain carbolfuchsin stain despite a washing with acid-alcohol. The bacteria are acquired in respiratory droplets and infect the lung tissues.

Tuberculosis is accompanied by the formation of **tubercles,** which are nodules on the lung tissue. The tubercle has a soft, cheeselike center and is surrounded by layers of macrophages and T-lymphocytes. When the lesions heal as calcified bodies, they are called **Ghon complexes.** In some individuals, the tubercles continue to grow, and the lesion may rupture to release microorganisms into the bloodstream for spread to other body organs. This condition is called **miliary tuberculosis.** Sometimes the disease is called **consumption.**

Tuberculosis may be treated over a period of months with several drugs, including isoniazid (INH), rifampin, streptomycin, pyrazinamide, ethambutol, and others. The **tuberculin skin test** is based on a type of cellular (delayed) hypersensitivity and is used to determine whether a person has had a previous exposure to tuberculosis antigens. One variation, the **Mantoux test,** uses dilutions of antigen called **PPD** (purified protein derivative), which are injected superficially to the skin to induce a reaction. A vaccine called **BCG (bacille Calmette Guerin)** is prepared from bovine tubercle bacilli and is available for immunization.

Pneumococcal pneumonia. Pneumococcal pneumonia is caused by *Streptococcus pneumoniae,* the **pneumococcus.** This organism is a Gram-positive pair of cocci occurring in chains. There are almost 100 serological types of the organism, and the vaccine currently available provides protection against approximately 25 of them. The disease involves the lung tissues and is accompanied by fever, consolidation of the lung (filling of the air spaces with bacteria, fluid, and debris), and severe chest pains, with blood in the sputum. Aggressive penicillin therapy is used in treatment. Many individuals are healthy carriers of the bacterium.

Mycoplasmal pneumonia. Mycoplasmal pneumonia is caused by *Mycoplasma pneumoniae,* a species of mycoplasma. Mycoplasmas are exceptionally small, submicroscopic bacteria (about 0.15 μm) that have no cell walls. Penicillin is therefore useless as a therapeutic agent. Most cases are accompanied by a mild pneumonia, and erythromycin is generally recommended for therapy. The disease is sometimes called **primary atypical pneumonia** and is often described as **walking pneumonia.**

Legionnaires' disease. Legionnaires' disease, or **legionellosis**, was first recognized in 1976 when an outbreak occurred among American Legion members attending a convention in Philadelphia. The causative agent is a Gram-negative rod called *Legionella pneumophila*. The organism exists where water collects and is airborne in wind gusts. Cases of Legionnaires' disease are accompanied by high fever, lung consolidation, and pneumonia. Erythromycin is used for therapy. A closely associated disease known as **Pontiac fever** is caused by the same organism.

Psittacosis. Psittacosis is caused by a species of chlamydia called *Chlamydia psittaci.* Chlamydiae are extremely tiny bacteria (0.25 μm) that cannot be seen with a light microscope. Psittacosis occurs in parrots, parakeets, and other psittacine birds, and when it is transferred to humans in airborne droplets, it manifests itself as a type of pneumonia with fever, headache, and lung consolidation. When the disease occurs in birds other than psittacines, it is known as **ornithosis.** Tetracyclines are effective drugs for therapy.

Chlamydial pneumonia. Chlamydial pneumonia is a recently recognized infection due to a species of chlamydia called *Chlamydia pneumoniae.* The infection resembles influenza and is treated successfully with tetracycline therapy.

Q fever. Q fever is due to a rickettsia known as *Coxiella burnetii.* The organism is transmitted by airborne droplets as well as by arthropods such as **ticks.** The infection resembles a form of pneumonia and is treated with tetracycline. Some cases are transmitted by contaminated or unpasteurized dairy products.

Viral Diseases

Common cold. Numerous viruses are capable of causing the syndrome known as the common cold. Among these are **rhinoviruses, coronaviruses,** and hundreds of strains of **adenoviruses.** Most cases are associated with sneezing, nasal discharge, congestion, coughing, and in some cases, middle ear infection. Therapies are directed at lessening the symptoms, and antiviral therapies are generally not available.

Influenza. The **influenza** virus consists of eight RNA strands, helically wound and enclosed in a capsid. The virus has an envelope with spikes containing **hemagglutinin (H)** and **neuraminidase (N).** Variations in the chemical character of the spikes account for the different forms and strains of influenza virus. The viruses are grouped as types A, B, and C and have names such as A(H2N4).

Influenza is accompanied by characteristic respiratory symptoms and muscle aches. Although the disease is rarely fatal, secondary bacterial infections may be a cause of death, and antibiotics may be given as precautionary measures. The drug amantadine has been found to lessen the symptoms of influenza, especially when used early in the infection.

RS virus disease. A serious form of viral pneumonia can be caused by the **respiratory syncytial virus.** This RNA virus causes cell cultures to fuse and form clusters called **syncytia** (singular, **syncytium**). In the human body, the virus causes severe coughing and wheezing, especially in children. Ribavirin may be administered to lessen the severity of disease.

Fungal and Protozoal Diseases

Histoplasmosis. **Histoplasmosis** is a fungal disease caused by the yeast *Histoplasma capsulatum.* In the body, the infection is similar to tuberculosis, especially in immunocompromised individuals. In severe cases, it may be a progressive disease that spreads to the other organs. Most cases are associated with bird and bat droppings. Amphotericin B is useful for therapy.

Blastomycosis. **Blastomycosis** is due to *Blastomyces dermatitidis,* a yeastlike fungus. The disease is found in regions of the Mississippi Valley and is spread in dust. Tuberculosislike lesions of the lung occur, and spread to other organs is possible. Amphotericin B is used for therapy.

Coccidioidomycosis. **Coccidioidomycosis** is due to *Coccidioides immitis,* a fungus found in the soil of the southwest United States. The disease is particularly prevalent in the San Joaquin Valley of California and is sometimes called **valley fever.** It is transmitted in dust, and its symptoms include fever, coughing, and general malaise. Progressive disease sometimes occurs. Farm workers are particularly disposed to the disease. Amphotericin B is used for therapy.

Aspergillosis. **Aspergillosis** is caused by *Aspergillus fumigatus,* a fungus. The fungus grows in the lung tissues and forms a compact ball of fungal mycelium, which blocks the respiratory passageways. Surgery is often needed to remove the mass of fungi.

***Pneumocystis* pneumonia.** ***Pneumocystis* pneumonia** is caused by *Pneumocystis carinii.* Although the organism is usually considered a protozoan, there is biochemical evidence that it may be a fungus. *Pneumocystis* pneumonia is associated with **AIDS** patients. The organisms grow in the lungs of immunocompromised individuals and cause severe consolidation, which may lead to death. A drug called pentamidine isethionate is valuable for therapy. Approximately half of the deaths associated with AIDS are due to *Pneumocystis* pneumonia. *Pneumocystis carinii* is a very complex organism with a life cycle involving mature cysts and highly resistant forms. It is present in the lungs of most individuals but does not invade the tissues unless the immune system has been compromised.

The digestive system consists of the gastrointestinal tract, which includes the oral cavity, pharynx, esophagus, stomach, and intestines, and a number of associated structures and glands such as the teeth, salivary glands, liver, and pancreas. These organs consume food, digest it, absorb nutrients, and eliminate waste that is not absorbed.

Bacterial Diseases

Dental caries. **Dental caries,** or cavities, is a universal microbiological problem. Most cases are caused by *Streptococcus mutans,* which adheres to the tooth enamel and produces glucans, which are a meshwork of glucose molecules. Together with bacteria and debris, glucans make up the **dental plaque.** The bacteria ferment carbohydrates in the diet and produce lactic acid, acetic acid, butyric acid, and other acids that damage the enamel. The susceptibility to tooth decay can be lessened by thorough brushing and flossing to remove *S. mutans* and by reducing the consumption of sugar.

Periodontal disease. **Periodontal disease** involves damage to the tissues surrounding and supporting the teeth. The gingiva, or gums, are also involved, as is the bony socket in which the tooth is embedded. Among the many causes of periodontal disease is *Bacteroides gingivalis,* an anaerobic, Gram-negative rod. Spirochetes such as species of *Treponema* also play a role.

Shigellosis. **Shigellosis** is also known as **bacillary dysentery.** It is caused by four species of the Gram-negative rod *Shigella: S. dysenteriae, S. boydii, S. sonnei,* and *S. flexneri.* Most cases occur in young children, and transmission takes place by an oral-fecal route. The disease is highly communicable and is initiated by a low number of bacteria as compared to other infections. The bacteria produce a powerful toxin (the **shigalike toxin**) that causes lesions and inflammation of the intestinal lining and stools streaked with blood and mucus. Dehydration is a threat, and rehydration is necessary to prevent death. Antimicrobial therapy is also available with a number of antibiotics, including quinolones.

Salmonellosis. Salmonellosis refers to a number of foodborne and waterborne infections due to species of *Salmonella.* The organisms are Gram-negative rods and include *S. enteritidis* and *S. choleraesuis.* They are transmitted by a fecal-oral route, and patients experience extensive diarrhea with fever, abdominal cramps, and nausea. The infection usually limits itself, and antibiotic therapy is not used unless severe complications exist. Chicken, egg, and poultry products are often involved because *Salmonella* strains live in domestic fowl.

Typhoid fever. Typhoid fever is caused by the Gram-negative, aerobic rod *Salmonella typhi.* The disease is transmitted by contaminated food and water and begins with a high fever lasting several days or weeks. A skin rash called **rose spots** is associated with the disease. Patients are tired, confused, and delirious, and the mortality rate without antibiotic therapy is high. Intestinal bleeding and wall perforation may occur. Chloramphenicol is used in therapy. The carrier state exists in people who have recovered. These people shed the bacteria in their feces and are a source of infection to other individuals.

Cholera. Cholera, caused by *Vibrio cholerae,* is a disease transmitted primarily by contaminated water. The etiologic agent is a short, curved, Gram-negative rod having a single polar flagellum. Its exotoxin binds to host cells, and the host epithelial cells secrete large quantities of chloride into the intestinal lumen followed by large amounts of water and sodium and other electrolytes. Massive diarrhea accompanies the disease, and dehydration often leads to death. The only effective treatment is rehydration accomplished by intravenous and oral rehydrating solutions.

***Escherichia coli* infections.** *Escherichia coli* is the Gram-negative rod routinely used in research and industrial microbiology because it is generally harmless. However, certain strains produce toxins or have the capability of invading tissue, and these strains can cause infections in humans. One disease attributed to *E. coli* is **traveler's diarrhea,** an infection developing in travelers to Caribbean and Central American countries, among others. **Infant diarrhea** and **urinary tract infections** are also caused by *E. coli*. *E. coli* **0157:H7** has been implicated in recent years in numerous foodborne outbreaks. Patients suffer hemorrhaging, especially in the kidneys, and infections can be serious.

Campylobacteriosis. Campylobacteriosis is caused by *Campylobacter jejuni,* a curved, Gram-negative rod often transmitted by contaminated milk. Patients experience bloody diarrhea, as well as abdominal pain and fever. Most infections limit themselves, but antibiotic therapy with erythromycin hastens recovery.

Gastric ulcers disease. In recent years, **gastric ulcers** have been related to the Gram-negative rod *Helicobacter pylori.* This organism survives in the lining of the stomach by producing enzymes to convert urea to ammonia, thereby raising the pH. Penetration of the stomach wall's mucosa follows. Antibiotics such as tetracycline have been used to limit the bacterium's proliferation.

Staphylococcal food poisoning. Staphylococcal food poisoning is the most frequently reported type of food poisoning in the United States. It is caused by toxin-producing strains of *Staphylococcus aureus.* The toxin, an **enterotoxin,** is produced in food and affects the gastrointestinal tract causing vomiting, diarrhea, and abdominal cramps. The incubation period is a short few hours, and the illness limits itself after a brief but intense period. Antibiotic therapy is not used. Fluid replacement may be necessary if severe diarrhea has taken place. Careful handling of foods, especially leftover foods, is paramount in preventing this disease.

Clostridial food poisoning. Clostridial food poisoning is due to *Clostridium perfringens,* a sporeforming, anaerobic rod. This organism produces its toxin in meat, and consumption of contaminated meat leads to mild gastroenteritis, with diarrhea. The infection is self-limiting and rarely requires antibiotic therapy. *Clostridium botulinum* also is transmitted in contaminated food. Its toxin affects the nervous system (botulism is discussed on page 182).

Leptospirosis. Leptospirosis is a disease of animals (such as dogs) as well as humans, where it causes damage to the liver and kidney. The etiologic agent is *Leptospira interrogans,* a spirochete. Humans usually become infected by contact with urine of the animals as the spirochete enters abrasions in the skin. Patients suffer muscle aches, fever, and infection of the liver. Kidney failure may also occur. Penicillin antibiotics are used for therapy.

Other bacterial diseases. A mild form of gastrointestinal illness is caused by *Vibrio parahaemolyticus.* This Gram-negative, curved rod often contaminates fish, and the diarrhea it causes may be mild or explosive. Low-grade fever, cramps, and vomiting accompany the illness. The organism lives in salt-water environments, especially in the region near Japan.

A type of colitis is caused by *Yersinia enterocolitica,* a Gram-negative rod that displays bipolar staining. This organism adheres to the epithelium of the intestine and produces an enterotoxin. Intense abdominal pain accompanies the infection. The organism is associated with leftover foods, especially those held in the refrigerator. Milk and animal products transmit the bacteria to humans.

A type of food poisoning is caused by *Bacillus cereus,* an aerobic, sporeforming rod. This organism's spores often survive the cooking process, and its toxins accumulate in vegetable and rice dishes. The infection is accompanied by vomiting or diarrhea or both.

Viral Diseases

Mumps. The virus that causes **mumps** contains RNA. Transmitted in saliva and respiratory secretions, it replicates in the host's respiratory tract and causes swelling of one or both of the **parotid glands** below the ear and near the angle of the jaw. Fever is sometimes present, and in adult males, complications may occur if the virus infects the testis. Inflammation of the testis is called **orchitis.** Immunity to mumps is rendered by an injection of the **MMR vaccine,** using attenuated mumps virus.

Hepatitis A. **Hepatitis A** is caused by an RNA virus usually placed in the Picornaviridae family. The virus passes among individuals by the fecal-oral route, and the disease is sometimes called **infectious hepatitis.** Individuals are contagious before they display symptoms and after symptoms have lessened. Contaminated food and water are often involved.

The hepatitis A virus affects the **liver.** Tissue damage is accompanied by vomiting, nausea, dark urine, and jaundice (a yellow discoloration of the skin and the whites of the eyes). Immunization may be rendered with an injection of the **hepatitis A vaccine** containing inactivated viruses. Prevention of symptoms is possible with hepatitis gamma globulin, a preparation of serum rich in hepatitis A antibodies. The hepatitis A virus is extremely resistant and remains active outside the body in the environment.

Hepatitis B. **Hepatitis B,** also called **serum hepatitis,** is caused by a DNA virus that is classified in the Hepadnaviridae. The virus is extremely fragile and passes directly from person to person, primarily in blood and semen. Hepatitis B is accompanied by liver infection, and in some cases, liver failure. Symptoms are similar to those in hepatitis A but tend to be more severe. Liver cancer (**hepatocarcinoma**) is a possible long-range complication of hepatitis B. Immunization may be rendered with an injection of genetically engineered

hepatitis B vaccine prepared in yeasts. Injections of gamma globulin containing hepatitis B antibodies are used for passive immunization in those infected by the virus.

Other forms of hepatitis. In addition to hepatitis A and hepatitis B, other forms of hepatitis are now known to exist. **Hepatitis C** is caused by an RNA virus transmitted by blood and semen. Most cases are associated with transfusions.

Delta hepatitis is related to an antigen called the **delta antigen,** which is a part of an RNA virus called the **delta virus.** Infection with this type of hepatitis accompanies infection with hepatitis B virus because the delta antigen relies on hepatitis B virus for its replication. This hepatitis is sometimes called **hepatitis D.**

Hepatitis E is also known to exist. The responsible virus is an RNA virus. Cases appear to be restricted to Asia, Africa, and India. Types of hepatitis such as these are often considered **non-A non-B hepatitis.**

Viral gastroenteritis. Viral gastroenteritis is a general expression for viral infection of the intestine. A major cause is the **rotavirus, a** virus transmitted by the fecal-oral route and capable of causing severe diarrhea. Dehydration may be a problem in patients, and antiviral therapies are generally inadequate.

Another possible cause of viral gastroenteritis is the **Norwalk agent,** probably a virus but not yet identified with certainty. The **Coxsackie virus** is an RNA virus also capable of causing intestinal infection. Contaminated food and water transmit this virus. Another possible cause of gastroenteritis is the **echovirus,** also an RNA virus.

Protozoal Diseases

Amoebic dysentery. Amoebic dysentery is caused by the amoeba *Entamoeba histolytica.* This protozoan exists in nature in the **cyst**

form and is transmitted by contaminated food and water. In patients, the amoebas revert to **trophozoites** (feeding forms) and invade the intestinal lining. Then they enter the bloodstream and move to distant organs, such as the liver and lung. Infected individuals pass the cysts in stools and remain carriers for long periods. A drug called metronidazole is used for therapy.

Giardiasis. Giardiasis is a protozoal disease caused by the flagellate *Giardia lamblia.* The organism is taken into the body in its cyst form in contaminated food and water. In the intestine, the trophozoites emerge from the cysts and multiply along the walls of the intestine. A foul-smelling, watery discharge accompanies the infection, followed by abdominal pain and diarrhea. Hikers, backpackers, and campers are particularly susceptible, since mountain streams often contain the cysts from wild animals. Metronidazole is used in therapy.

Balantidiasis. Balantidiasis is caused by the protozoal ciliate *Balantidium coli.* This protozoan enters the body as a cyst, and the trophozoite form emerges in the intestine. Tissue invasion may occur, and diarrhea is accompanied by blood and pus in the stools. Symptoms tend to last for long periods. Patients become carriers. Metronidazole can be used in therapy.

Cryptosporidiosis. Cryptosporidiosis is caused by species of *Cryptosporidium* such as *C. parvum* and *C. coccidi.* The organism invades the intestinal epithelium and induces mild gastroenteritis with abdominal pain and watery diarrhea. Symptoms tend to be very severe in AIDS patients, and the massive diarrhea can be lethal. No treatments are known as of this writing. Water is believed to be the main mode of transmission. Many methods for purifying water permit this organism to pass, and modifications of these treatment methods are now being considered.

Parasitic Diseases

Parasitic diseases of the digestive system usually involve worms, also known as **helminths.** In most cases, the worms multiply in the system, and when the worm burden becomes high, the symptoms of disease ensue. Poor sanitation contributes to the occurrence of parasitic (helminthic) infections.

Pinworm disease. **Pinworm disease** is caused by the small roundworm *Enterobius vermicularis.* Infections occur after ingestion of pinworm eggs. The eggs hatch, and adult females lay their eggs near the body surface, particularly near the anus. Young children are usually those infected. Several drugs are available for treating pinworm infection.

Roundworm disease. **Roundworm disease** is due to *Ascaris lumbricoides.* The infection begins with the ingestion of roundworm eggs, which yield roundworms that burrow through the intestinal wall to the bloodstream, ultimately reaching the lungs. The worms reenter the digestive system when they are coughed up from the lungs and swallowed. A large number of eggs cause respiratory distress, and intestinal obstruction may also develop due to heavy worm burdens.

Hookworm disease. **Hookworm disease** may be caused by either of two species of roundworms: *Ancylostoma duodenale* (the Old-World hookworm) or *Necator americanus* (the New-World hookworm). The larvae of the hookworm penetrate the human skin, usually through the foot, and the hookworms pass through the bloodstream to the lungs, from where they are coughed up and swallowed to the digestive system. The worms use their hooks to hold fast to the intestinal lining. Then they suck the blood and multiply. Infestations lead to anemia, with much fatigue and weakness. Hookworm disease is common where people go barefoot.

Strongyloidiasis. **Strongyloidiasis** is caused by the roundworm *Strongyloides stercoralis.* The worms penetrate the human skin and pass from the blood to the lungs, and eventually to the digestive system. Infestations result in high worm burdens and intestinal blockages. Invasion of the intestinal wall may accompany the disease, especially in immunocompromised individuals.

Whipworm disease. **Whipworm disease** is caused by the roundworm *Trichuris trichiura,* called the whipworm because its body resembles a whip. Eggs are ingested in food and water, and they hatch in the digestive tract to become adults. The adults lay their eggs, which are passed in the feces. Heavy worm burdens in the intestine cause irritation, inflammation, and other symptoms of obstruction.

Trichinosis. **Trichinosis** is due to the roundworm *Trichinella spiralis.* This parasite infects the muscle tissues of pigs and is usually passed to humans by improperly cooked pork products. The worms enter the human bloodstream from the intestine and form cysts in the muscles. Heavy worm burdens in the rib muscles cause severe pain. The worms also migrate to the heart muscle, diaphragm, and lungs. Proper cooking of pork products is paramount in preventing infection.

Taenisias. **Taenisias** is caused by a **tapeworm,** which is a type of flatworm. Two tapeworms are important in humans: *Taenia solium,* the pork tapeworm, and *Taenia saginata,* the beef tapeworm. Humans are infected when they eat contaminated pork or beef, respectively. Adult worms attach to the intestinal lining using their sucker devices and hooks. As the tapeworm lengthens, it adds segments called **proglottids.** Eventually the worm may be several feet long. Proglottids break free and are released in the feces to infect pigs or cattle that feed in the soil. Heavy worm burdens may cause intestinal blockage and abdominal pain.

Hydatid disease. Hydatid disease is caused by a type of small flatworm called a tapeworm. The tapeworm involved is *Echinococcus granulosus.* Humans are infected by contact with animal feces (especially, that of canines), and the worms form **hydatid cysts** in the tissues. The large cysts cause damage to organs such as the liver or lung.

Liver fluke disease. The liver can be infected by a leaflike flatworm known as a **fluke. Liver fluke disease** is due to the **sheep liver fluke** known as *Fasciola hepatica,* or it may be caused by the **Chinese liver fluke** referred to as *Clonorchis sinensis.* The flukes are ingested with water plants such as watercress. The worm larvae migrate to the liver where they develop into adults. Liver damage and jaundice accompany the disease. Outside the body, the flukes live in **snails,** the intermediary hosts.

The reproductive systems of males and females open to the external environment, and therefore, the organs can be easily reached by infectious organisms. The diseases may then spread to deeper organs of the human body.

Bacterial Diseases

Gonorrhea. At this writing, **gonorrhea** is the most-reported infectious disease in the United States. The etiologic agent is the Gram-negative diplococcus *Neisseria gonorrhoeae.* The organism attaches to the epithelial cells of the male and female urethra causing **urethritis.** Transmission occurs during sexual contact, and males exhibit more extensive symptoms than do females, with pain on urination and a whitish discharge from the urethra. Treatment with tetracycline, penicillin, and other antibiotics is usually successful.

Complications of gonorrhea may involve many organs. For example, in females, the Fallopian tubes may be blocked with scar tissue, thereby preventing passage of the egg cells and resulting in sterility. A similar complication may occur when the epididymis and vas deferens are blocked in males. Many females suffer **pelvic inflammatory disease (PID)**, an inflammation of organs of the pelvic cavity such as the uterus, cervix, and ovaries. Infection may also occur in the rectum, pharynx, meninges, and joints. Newborns subjected to *N. gonorrhoeae* during passage through the birth canal may suffer eye infection called **gonococcal ophthalmia.** Treatment with silver nitrate and/or erythromycin shortly after birth prevents infection.

Chlamydia. A gonorrhealike infection called **chlamydia** is caused by *Chlamydia trachomatis,* a member of the chlamydia group of bacteria. The disease is often referred to as **nongonococcal urethritis** to distinguish it from gonorrhea. It is accompanied by pain during urination, a frequent desire to urinate, and a watery discharge. Several million people are believed to suffer from it annually. Tetracycline is used in therapy. Pelvic inflammatory disease may complicate the condition. Sterility is also a long-term complication. **Chlamydial ophthalmia** may occur in the eyes of newborns.

Mycoplasmal and ureaplasmal urethritis. **Mycoplasmal urethritis** is caused by a mycoplasma known as *Mycoplasma hominis,* while **ureaplasmal urethritis** is due to a mycoplasma known as *Ureaplasma urealyticum.* Both organisms cause infection of the urethra, with symptoms similar to those of gonorrhea and chlamydia. Tetracycline is used to treat both conditions, and PID may complicate the condition.

Syphilis. **Syphilis** has been known to exist for many centuries and was once known as the **Great Pox.** It is caused by the spirochete *Treponema pallidum.* Transmitted by sexual contact, the etiologic agent causes a disease occurring in three stages. The **primary stage** is accompanied by the **chancre,** a raised, hard, dry, crusty sore occurring at the site of infection. Spirochetes observed from the chancre constitute diagnosis. Penicillin therapy at this stage is successful.

The **secondary stage** of syphilis occurs several weeks after the chancre disappears. This stage is accompanied by an influenzalike syndrome of the respiratory system, a skin rash over the body surface with spirochete-laden lesions (pox), loss of hair, and mild fever. Treatment continues to be successful at this stage. A **latent period** follows, and in a small percentage of cases, the disease recurs in the **tertiary stage.** This stage is probably an immunological reaction. It is characterized by gummy, rubbery masses of damaged tissues called **gummas** occurring in the nervous and cardiovascular systems. In the most severe cases, aneurysms and paralysis may develop and mental deficiencies may become severe. Treatment at this stage is not always successful.

Congenital syphilis may occur if spirochetes pass between a pregnant woman and her fetus. Numerous diagnostic tests exist for the detection of both spirochetes and antibodies produced against the spirochetes.

Chancroid. Infection of the reproductive tract may be due to *Haemophilus ducreyi.* This small, Gram-negative rod causes an STD called **chancroid.** The disease is characterized by a swollen, painful ulcer on the genital organs, with infection of the lymph nodes called **buboes.** It is referred to as **soft chancre** and is treated with tetracycline. Sexual contact is the mode of transmission.

Vaginitis. Another disease of bacterial origin is **vaginitis** due to *Gardnerella vaginalis.* The bacterium is a Gram-negative rod commonly found in the vagina as an opportunistic organism. Often the infection is associated with the destruction of lactobacilli normally found in the vaginal tract (such as by excessive antibiotic use). The drug metronidazole is used in therapy.

Lymphogranuloma venereum. Lymphogranuloma venereum is caused by a strain of *Chlamydia trachomatis,* the organism that also causes chlamydia. The disease is characterized by lesions at the infection site followed by swollen lymph nodes. Transmission occurs during sexual contact. Tetracycline is used for therapy.

Viral Diseases

Genital herpes. The **herpes simplex virus** is responsible for cases of **genital herpes.** The virus is a DNA icosahedral virion, the same virus that causes cold sores of the mouth. However, the strain of virus is usually type II in genital herpes (type I in cold sores). Painful urination accompanies the disease, and fluid-filled vesicles occur on the genital organs. Recurrences occur many times, but their frequency and severity can be limited by treatment with acyclovir. Transmission to the newborn can occur during birth. The virus is also capable of crossing the placenta and affecting the fetus before birth.

Genital warts. **Genital warts** is considered a viral disease. Most cases are due to **papillomaviruses,** which have DNA. Warts may be smooth and flat or large with fingerlike projections. Often the condition is called **condyloma acuminatum.** Cases of **cervical cancer** have been related to genital warts in women. Treatments often consist of excision of the wart.

Fungal and Protozoal Diseases

Candidiasis (yeast disease). **Candidiasis** is a fungal disease of the reproductive tract caused by the yeast *Candida albicans.* Infections usually accompany destruction of the local population of bacteria, often related to overuse of antibiotics. Cases of candidiasis are accompanied by lesions similar to those in thrush, as well as severe pruritis, and a yellowish, cheesy discharge. Diagnosis is performed by visual observation of the yeasts. Treatment with a number of antifungal drugs is recommended, including nystatin, clotrimazole, and miconazole.

Trichomoniasis. The only protozoal disease of the reproductive tract is **trichomoniasis,** due to the flagellate *Trichomonas vaginalis.* This organism grows along the mucosa of the reproductive tract, causing internal discomfort and a profuse, green-yellow discharge with a foul odor. The organism is observed in urine and discharge specimens. Metronidazole is used for successful therapy.

Human activities generate a tremendous volume of sewage and wastewater that require treatment before discharge into waterways. Often this wastewater contains excessive amounts of nitrogen, phosphorus, and metal compounds, as well as organic pollutants that would overwhelm waterways with an unreasonable burden. Wastewater also contains chemical wastes that are not biodegradable, as well as pathogenic microorganisms that can cause infectious disease.

Sewage and Wastewater Treatment

The chemical and biological waste in sewage and water must be broken down before it is deposited to the soil and environment. This breakdown can effectively be controlled by managing the microbial population in waters and encouraging microorganisms to digest the organic matter. The water must then be purified before it is considered fit to drink. Water taken from ground sources must also be treated before consumption.

Water purification. To purify water for drinking, a number of processes are conducted to reduce the microbial population and maintain that population at a safe level. First, the solid matter is allowed to settle out in a **sedimentation tank.** Flocculating materials such as alum are used to drag microorganisms to the bottom of the tank.

Then the filtration process is begun. Water is filtered through either a **slow sand filter** or a **rapid sand filter.** These processes remove 99 percent of the microorganisms. The slow sand filter is composed of finer grains of sand, and the filtration process takes longer than in the rapid sand filter, where larger grains are used.

Many communities then purify the water by **chlorination.** When added to water, chlorine maintains the low microbial count and ensures that the water remains safe for drinking purposes. Chlorine gas or hypochlorite (NaOCl) is used for chlorination purposes. The water is chlorinated until a slight residue of chlorine remains.

Sewage treatment. Sewage treatment involves a more complex set of procedures than are needed for water purification because the volume of organic matter and the variety of microorganisms are much greater.

The first treatment, or **primary treatment,** of sewage and wastewater involves the removal in settling tanks of particulate matter such as plant waste. The solids that sediment are strained off, and the sludge is collected to be burned or buried in landfills. Alternatively, it can be treated in an anaerobic sludge-digesting tank, as follows.

During the **secondary treatment** of wastewater and sewage, the microbial population of liquid and sludge waste is reduced. In the **anaerobic** sludge digester, microorganisms break down the organic matter of proteins, lipids, and cellulose into smaller substances for metabolism by other organisms. Results of these breakdowns include organic acids, alcohols, and simple compounds. Methane gas is produced in the sludge tank, and it can be burned as a fuel to operate the waste treatment facility. The remaining sludge is incinerated or buried in a landfill, and its fluid is recycled and purified (Figure 35).

In **aerobic** secondary sewage treatment, the fluid waste is aerated and then passed through a **trickling filter.** In this process, the liquid waste is sprayed over a bed of crushed rocks, tree bark, or other filtering material. Colonies of bacteria, fungi, and protozoa grow in the bed and act as secondary filters to remove organic materials. The microorganisms metabolize organic compounds and convert them to carbon dioxide, sulfate, phosphates, nitrates, and other ions. The material that comes through the filter has been 99 percent cleansed of microorganisms.

Liquid waste can also be treated in an **activated digester** after it has been vigorously aerated. Slime-forming bacteria form masses that trap other microorganisms to remove them from the water. Treatment for several hours reduces the microbial population significantly, and the clear fluid is removed for purification. The sludge is placed in a landfill or at sea.

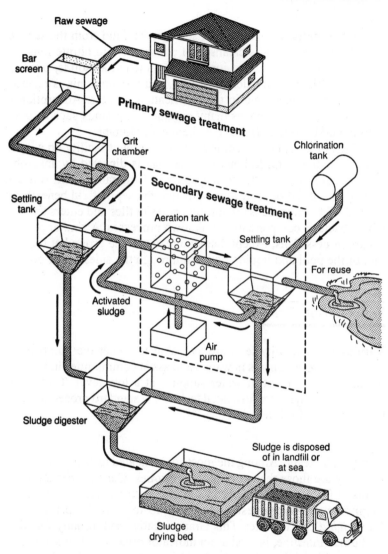

A view of the methods used in sewage treatment in a large municipality. Primary treatment is represented by the steps preceding secondary treatment, and tertiary treatment is performed in the chlorination tank at the conclusion of the process.

■ Figure 35 ■

In the **tertiary treatment** of sewage, the fluid from the secondary treatment process is cleansed of phosphate and nitrate ions that might cause pollution. The ions are precipitated as solids, often by combining them with calcium or iron, and the ammonia is released by oxidizing it to nitrate in the nitrification process. Adsorption to activated charcoal removes many organic compounds such as polychlorinated biphenyls (PCBs), a chemical pollutant.

The home septic system is a waste treatment facility on a small scale. In a **septic tank,** household sewage is digested by anaerobic bacteria, and solids settle to the bottom of the tank. Solid waste is carried out of the outflow apparatus into the septic field beneath the ground. The water seeps out through holes in tiles and enters the soil, where bacteria complete the breakdown processes. A similar process occurs in **cesspools,** except that sludge enters the ground at the bottom of the pool and liquids flow out through the sides of the pool.

Water Quality Tests

Various water quality tests are available to detect the number and types of microorganisms in waters and assist communities in keeping the microbial content of water supplies at a low level. These tests vary from the more sophisticated tests to the standard procedures that have been used for decades.

Gene probe tests. Among the most sophisticated tests for water bacteriology are those that employ **gene probes.** Gene probes are fragments of DNA that seek out and combine with complementary DNA fragments. Often the test is designed to test for the presence of *Escherichia coli* in water. This Gram-negative rod, usually found in the human intestine, is used as an indicator organism. If it is present, then it is likely that the water has been contaminated with human feces. The feces may contain microbial pathogens.

To use a gene probe test for *E. coli* in water, the water is treated to disrupt any bacteria present and release their nucleic acid. Then a specific *E. coli* probe is added to the water. Like a left hand seeking a right hand, the probe searches through all the nucleic acid in the water and unites with the *E. coli* DNA, if present. A radioactive signal indicates that a match has been made. If no radioactivity is emitted, then the gene probe has been unable to locate its matching DNA, and *E. coli* is probably absent from the water.

The membrane filter technique. The **membrane filter technique** uses a filtration apparatus and a cellulose filter called a membrane filter. A 100-ml sample of water is passed through the filter, and the filter pad is then transferred to a bacteriological growth medium. Bacteria trapped in the filter grow on the medium and form colonies. By counting the colonies, an estimate can be made of the number of bacteria in the original 100-ml sample.

The standard plate count. It is generally impractical to test for all pathogenic organisms, but the total number of bacteria can be calculated. One test is the **standard plate count.** In this test, samples of water are diluted in jars containing 99-ml sterile water, and samples are placed in Petri dishes with **nutrient agar** or other nutritious medium (Figure 36). After incubation, the colony count is taken and multiplied by the dilution factor to obtain the total number of bacteria per ml of sample.

Indicator bacteria can be detected to give an estimate of pathogens. The most common indicator organisms in water bacteriology are the **coliform bacteria.** These are Gram-negative rods normally found in the intestine and typified by *Escherichia coli*. To test for the presence of coliforms, a standard plate count can be performed, with **violet red bile agar** used as the growth medium to encourage proliferation of the coliform bacteria.

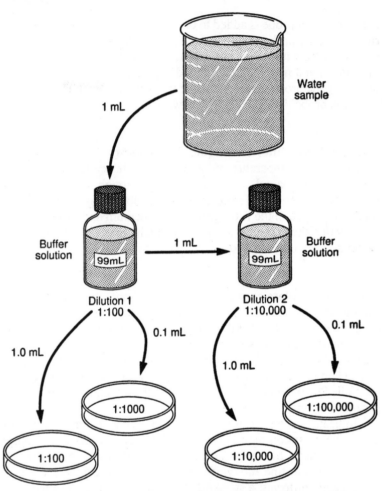

The standard plate count procedure. A 1-mL water sample is diluted in buffer solution, and various amounts are placed with nutrient medium into Petri dishes to encourage bacterial colonies to form. The colony count is multiplied by the dilution factor to yield the total plate count.

■ Figure 36 ■

The most probable number test. In the **most probable number (MPN) test,** tubes of lactose broth are inoculated with water samples measuring 10 ml, 1 ml, and 0.1 ml. During incubation, coliform organisms produce gas. Depending upon which tubes from which water samples display gas, an MPN table is consulted and a statistical range of the number of coliform bacteria is determined. The MPN test is very easy to perform and interpret, but it does not determine the exact number of bacteria as the standard plate count does.

In order to test for the presence of *E. coli* in water, a medium called **eosin methylene blue (EMB) agar** is used. On this medium, *E. coli* colonies become green with a metallic fluorescent sheen.

The biochemical oxygen demand (BOD). The extent of pollution in wastewater can be determined by measuring the **biochemical oxygen demand (BOD).** The BOD is the amount of oxygen required by the microorganisms during their growth in wastewater.

The BOD test is begun by noting the oxygen concentration in a sample of water before incubation. The water is then incubated in an air-tight, stoppered bottle for a period of about five days. A temperature of between 5° and 20°C is used. The oxygen concentration in the water is then noted again, and the difference in the dissolved oxygen is the BOD. A higher BOD indicates presence of a higher amount of organic matter. High BOD values are found in wastewater from agricultural communities, food processing plants, and certain industries.

In the biosphere, microorganisms play a key role in maintaining the chemical balance of available nutrients and metabolic waste products. In this way, they help preserve the natural environment. In addition, microorganisms are important in the elemental cycles occurring in the soil.

An **ecosystem** is the community of organisms found in a physically defined space. In the soil ecosystem, microorganisms are the largest contributors of organic matter. This organic matter is derived from the metabolism of animal and plant waste. The decay-resistant organic matter not recycled combines with mineral particles to form the dark-colored material of soil called **humus.** Humus increases the soil's ability to retain air and water.

The Nitrogen Cycle

Renewable resources can be recycled for reuse through the interactions of natural processes of metabolism. Microorganisms are essential in the webs of metabolic activities that renew the earth's natural resources. Among the most important biogeochemical cycles is the **nitrogen cycle.**

Nitrogen is a key cellular element of amino acids, purines, pyrimidines, and certain coenzymes. The element accounts for about 9 to 15 percent of the dry weight of a cell. Proteins and other organic compounds of life could not be formed without nitrogen.

Ammonification. In the nitrogen cycle, many organisms obtain their nitrogen from organic sources such as amino acids or purines, while others obtain it from inorganic compounds such as nitrogen gas (N_2), ammonia (NH_3), or nitrate (NO_3^{-1}). Before nitrate or nitrogen gas can be used, however, the nitrogen in the compounds must be changed into ammonia, a process called **ammonification.** The ammonia is then brought into the living system by an enzyme-catalyzed pathway in which glutamic acid and glutamine form. These amino acids are then used to synthesize other nitrogen compounds in the cell (Figure 37).

Nitrogen fixation. The principal reservoir of nitrogen on earth is the atmosphere, which contains about 80 percent nitrogen. In the process of **nitrogen fixation,** nitrogen gas from the atmosphere is used to form ammonia by the chemical process of reduction. Nitrogen fixation is performed by free-living bacteria as well as by bacteria growing in **symbiosis** with leguminous plants (plants that bear their seeds in pods, such as peas, beans, alfalfa, clover, and soybeans).

Nitrogen fixation is accomplished by species of *Rhizobium* inhabiting the roots of leguminous plants in a mutually beneficial (symbiotic) relationship. These Gram-negative bacteria penetrate the root hairs and form an infection thread that becomes a **root nodule.** Here the bacteria fix atmospheric nitrogen, while deriving nutrients from the plant.

There are many genera of free-living bacteria that exist apart from legumes and fix nitrogen in the soil. Among the important ones are species of *Azotobacter, Azospirillum, Bacillus, Beijerinckia,* and numerous species of cyanobacteria.

Once nitrogen has been incorporated into ammonia, the ammonia is used for various organic substances. Later, when plants, animals, and microorganisms die, the nitrogen is recycled by forming ammonia once again in the process of ammonification. For example, proteins and nucleic acids are broken down first to amino acids and purines and then to acids, gases, and ammonia. Ammonification also occurs from animal excretory products such as urea, the major component of the urine. The urea is broken down by urea-digesting bacteria, and ammonia is released.

Nitrification. The conversion of ammonia to nitrate (NO_3^{-1}) is the process of **nitrification.** Nitrifying bacteria, such as species of *Nitrosomonas* and *Nitrosococcus,* are involved. *Nitrosomonas* spe-

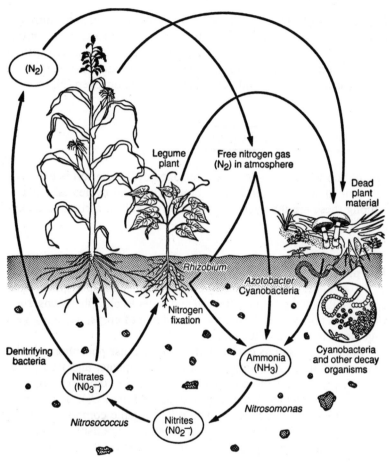

The complex interactions of the nitrogen cycle as it occurs in the soil.

■ Figure 37 ■

cies convert ammonia to **nitrite** (NO_2^{-1}); then ***Nitrosococcus*** species convert the nitrite to **nitrate** (NO_3^{-1}). Nitrification occurs in soils, fresh water, and marine environments. The nitrate that results serves as an important nitrogen source for plants.

Denitrification. **Denitrification** is the process in which the nitrogen of nitrate is released as gaseous nitrogen. This process makes nitrogen available to bacteria that use it for nitrogen fixation. Denitrification is accomplished by numerous bacteria that reduce nitrite (NO_2^{-1}) to nitrous oxide (N_2O) and then to atmospheric nitrogen (N_2).

Other Biogeochemical Cycles

In addition to being a site for the nitrogen cycle, the soil is the environment in which several other biogeochemical cycles take place. Among these are the cycles of phosphorus, sulfur, carbon, and oxygen.

The phosphorus cycle. Living things use **phosphorus** compounds in the synthesis of nucleotides, phospholipids, and phosphorylated proteins. Phosphorus enters the soil and water as phosphate ions, such as calcium phosphate, during the breakdown of crops, decaying garbage, leaf litter, and other sources.

In the **phosphorus cycle,** microorganisms use phosphorus in the form of calcium phosphate, magnesium phosphate, and iron phosphate. They release the phosphorus from these complexes and assimilate the phosphorus as the phosphate ion (PO_4). This ion is incorporated into DNA, RNA, and other organic compounds using phosphate, including phospholipids. When the organisms are used as foods by larger organisms, the phosphorus enters and is concentrated in the food chain.

The sulfur cycle. **Sulfur** makes up a small percentage of the dry weight of a cell (approximately 1 percent), but it is an important element in the formation of certain amino acids such as cystine, methionine, and glutathione. It is also used in the formation of many enzymes.

Many bacteria have an important place in the **sulfur cycle** in the soil. Sulfate-reducing bacteria grow in mud and anaerobic water environments, where they reduce sulfur-containing amino acids to hydrogen sulfide (H_2S). **Hydrogen sulfide** accumulates in the mud of a swamp and gives the environment an odor of rotten eggs.

In the cycle's next step, photosynthetic sulfur bacteria metabolize the hydrogen sulfide anaerobically. They oxidize the H_2S, thereby releasing the sulfur as elemental sulfur (S). **Elemental sulfur** accumulates in the soil. Species of colorless sulfur bacteria, including members of the genera *Thiobacillus, Beggiatoa,* and *Thiothrix,* also metabolize the hydrogen sulfide, converting it to **sulfate ions,** which are then made available to plants for amino acid formation.

The carbon cycle. Most of the organic matter present in soil originates in plant material from dead leaves, rotting trees, decaying roots, and other plant tissues. Animal tissues enter the soil after death. In the **carbon cycle,** soil bacteria and fungi recycle the **carbon** of proteins, fats, and carbohydrates by using the organic plant and animal matter in their metabolism. Without the recycling of carbon, life would suffer an irreversible decline as nutrients essential for life were bound in complex molecules.

The organic matter of organisms is digested by extracellular microbial enzymes into soluble products. Fungi and bacteria then metabolize the soluble organic products to simpler products such as carbon dioxide and acetic, propionic, and other small acids, as well as other materials available for plant growth. These elements are made available to the root systems of plants. Undigested plant and animal matter becomes part of the humus.

The oxygen cycle. In the **oxygen cycle, oxygen** is a key element for the chemical reactions of cellular respiration (glycolysis, Krebs cycle, electron transport, chemiosmosis). The atmosphere is the chief reservoir of oxygen available for these processes. Oxygen is returned to the atmosphere for use in metabolism by photosynthetic green plants and photosynthetic microorganisms such as cyanobacteria. During the process of photosynthesis, these organisms liberate oxygen from water and release it to the atmosphere. The oxygen is then available to heterotrophic organisms for use in their metabolism.

Food is considered contaminated when unwanted microorganisms are present. Most of the time the contamination is natural, but sometimes it is artificial. **Natural contamination** occurs when microorganisms attach themselves to foods while the foods are in their growing stages. For instance, fruits are often contaminated with yeasts because yeasts ferment the carbohydrates in fruits. **Artificial contamination** occurs when food is handled or processed, such as when fecal bacteria enter food through improper handling procedures.

Food Spoilage

Food spoilage is a disagreeable change or departure from the food's normal state. Such a change can be detected with the senses of smell, taste, touch, or vision. Changes occurring in food depend upon the composition of food and the microorganisms present in it and result from chemical reactions relating to the metabolic activities of microorganisms as they grow in the food.

Types of spoilage. Various physical, chemical, and biological factors play contributing roles in spoilage. For instance, microorganisms that break down fats grow in **sweet butter** (unsalted butter) and cause a type of spoilage called **rancidity.** Certain types of fungi and bacteria fall into this category. Species of the Gram-negative bacterial rod *Pseudomonas* are major causes of rancidity. The microorganisms break down the fats in butter to produce glycerol and acids, both of which are responsible for the smell and taste of rancid butter.

Another example occurs in **meat,** which is primarily protein. Bacteria able to digest protein (proteolytic bacteria) break down the protein in meat and release odoriferous products such as putrescine and cadaverine. Chemical products such as these result from the incomplete utilization of the amino acids in the protein.

Food spoilage can also result in a sour taste. If **milk** is kept too long, for example, it will sour. In this case, bacteria that have survived pasteurization grow in the milk and produce acid from the carbohydrate lactose in it. The spoilage will occur more rapidly if the milk is held at room temperature than if refrigerated. The sour taste is due to the presence of lactic acid, acetic acid, butyric acid, and other food acids.

Sources of microorganisms. The general sources of food spoilage microorganisms are the air, soil, sewage, and animal wastes. Microorganisms clinging to foods grown in the ground are potential spoilers of the food. Meats and fish products are contaminated by bacteria from the animal's internal organs, skin, and feet. **Meat** is rapidly contaminated when it is ground for hamburger or sausage because the bacteria normally present on the outside of the meat move into the chopped meat where there are many air pockets and a rich supply of moisture. **Fish tissues** are contaminated more readily than meat because they are of a looser consistency and are easily penetrated.

Canned foods are sterilized before being placed on the grocery shelf, but if the sterilization has been unsuccessful, contamination or food spoilage may occur. Swollen cans usually contain **gas** produced by members of the genus *Clostridium.* Sour spoilage without gas is commonly due to members of the genus *Bacillus.* This type of spoilage is called **flat-sour spoilage.** Lactobacilli are responsible for **acid spoilage** when they break down the carbohydrates in foods and produce detectable amounts of acid.

Among the important criteria determining the type of spoilage are the nature of the food preserved, the length of time before it is consumed, and the handling methods needed to process the foods. Various criteria determine which preservation methods are used.

Food Preservation

Food preservation methods are intended to keep microorganisms out of foods, remove microorganisms from contaminated foods, and hinder the growth and activity of microorganisms already in foods.

To keep microorganisms out of food, contamination is minimized during the entire food preparation process by sterilizing equipment, sanitizing it, and sealing products in wrapping materials. Microorganisms may be removed from liquid foods by **filtering** and sedimenting them or by washing and trimming them. **Washing** is particularly valuable for vegetables and fruits, and **trimming** is useful for meats and poultry products.

Heat. When **heat** is used to preserve foods, the number of microorganisms present, the **microbial load,** is an important consideration. Various types of microorganisms must also be considered because different levels of resistance exist. For example, bacterial spores are much more difficult to kill than vegetative bacilli. In addition, increasing acidity enhances the killing process in food preservation.

Three basic heat treatments are used in food preservation: **pasteurization,** in which foods are treated at about 62°C for 30 minutes or 72°C for 15 to 17 seconds; **hot filling,** in which liquid foods and juices are boiled before being placed into containers; and **steam treatment** under pressure, such as used in the canning method. Each food preserved must be studied to determine how long it takes to kill the most resistant organisms present. The heat resistance of microorganisms is usually expressed as the **thermal death time,** the time necessary at a certain temperature to kill a stated number of particular microorganisms under specified conditions.

In the **canning** process, the product is washed to remove soil. It is then blanched by a short period of exposure to hot water to deactivate enzymes in the food. Diseased sections in the food are removed, and the food is placed into cans by a filling machine. Sealed cans are then placed into a sterilizing machine called a **retort,** and the food is processed for a designated time and temperature.

Low temperatures. Low temperatures are used to preserve food by lowering microbial activity through the reduction of microbial enzymes. However, psychrophilic bacteria are known to grow even at cold **refrigerator** temperatures. These bacteria include members of the genera *Pseudomonas, Alcaligenes, Micrococcus,* and *Flavobacterium.* Fungi also grow at refrigeration temperatures.

Slow freezing and quick freezing are used for long-term preservation. **Freezing** reduces the number of microorganisms in foods but does not kill them all. In microorganisms, cell proteins undergo denaturation due to increasing concentrations of solutes in the unfrozen water in foods, and damage is caused by ice crystals.

Chemicals. Several kinds of chemicals can be used for food preservation, including **propionic acid, sorbic acid, benzoic acid,** and **sulfur dioxide.** These acids are acceptable because they can be metabolized by the human body. Some **antibiotics** can also be used, depending upon local laws and ordinances. Tetracycline, for example, is often used to preserve meats. Storage and cooking normally eliminates the last remnants of antibiotic.

In many foods, the **natural acids** act as preservatives. In sauerkraut, for example, lactic acid and acetic acid prevent contamination, while in fermented milks (yogurt, sour cream), acids perform the same function. For centuries, foods were prepared in this manner as a way of preventing microbial spoilage.

Drying. Drying is used to preserve food by placing foods in the sun and permitting the water to evaporate. **Belt, tunnel,** and **cabinet dryers** are used in industry for such things as instant coffee and cocoa. Freeze-drying, a process called **lyophilization,** is also valuable for producing a product free of moisture and very light.

Radiations. **Ultraviolet radiation** is valuable for reducing surface contamination on several foods. This short-wavelength light has been used in the cold storage units of meat processing plants. Ionizing radiations such as **gamma rays** can be used to preserve certain types of vegetables, fruits, and spices, according to state and U.S. federal regulations.

Microbial Determinations

To assess the presence and extent of microbial contamination in food, it is standard practice to perform several types of **microbial determinations.** These determinations are important because microorganisms from foods can cause such diseases as staphylococcal food poisoning, salmonellosis, typhoid fever, cholera, and gastroenteritis.

The standard plate count. One method for determining the number of bacteria in foods is the **standard plate count.** The procedure is performed by taking a gram of food sample and diluting it in 99-ml bottles of sterile water or buffer in the method described in the chapter on aquatic microbiology. The number of bacterial colonies is multiplied by the dilution factor to determine the number of bacteria per gram of food.

Coliform and fungal determinations. In food, it is valuable to assess the number of **coliform bacteria.** These Gram-negative intestinal rods do not cause disease but are valuable indicators of fecal contamination. Presumably, when coliform bacteria are present, the food has been contaminated with fecal matter and is unfit to consume. The coliform most commonly sought is *Escherichia coli.* Various types of bacteriological media, such as violet red bile agar, are available for cultivating this bacteria, and the standard plate count technique can be performed to assess the number of coliform bacteria per gram of food.

Fungi can be assessed in foods by using a medium such as Sabouraud dextrose agar. This medium encourages fungal spores to germinate and form visible masses of filaments called mycelia. A count of the resulting mycelia gives an estimate of the fungal contamination of the food.

The phosphatase test. To determine contamination in milk, the plate count technique can be used, but a more rapid test is the **phosphatase test** (Figure 38). Phosphatase is an enzyme destroyed by the pasteurization process. However, if the test for phosphatase shows that it is present after the milk has been treated, then the pasteurization has been unsuccessful. Testing for phosphatase is a more rapid and efficient method for determining contamination than the plate count technique.

Foods from Microorganisms

Microorganisms are widely used in the food industry to produce various types of foods that are both nutritious and preserved from spoilage because of their acid content.

Dairy foods. In the dairy industry, many products result from fermentation by microorganisms in milk and the products of milk. For example, **buttermilk** results from the souring of low-fat milk by lactic acid. The flavor is due to substances such as diacetyl and acetaldehyde, which are produced by species of *Streptococcus, Leuconostoc,* and *Lactobacillus* as they grow.

A fermented milk product with a puddinglike consistency is **yogurt.** Two bacteria, *Streptococcus thermophilus* and *Lactobacillus bulgaricus,* are essential to its production. After the milk has been heated to achieve evaporation, the bacteria are added, and the condensed milk is set aside at a warm temperature to produce the yogurt. **Sour cream** is produced in a similar way, using cream as a starter material.

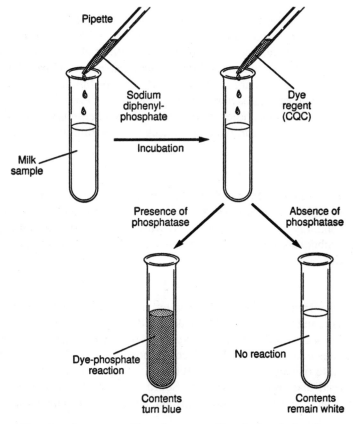

Pipette

Sodium
diphenyl-
phosphate

Dye
regent
(CQC)

Milk
sample

Incubation

Presence of
phosphatase

Absence of
phosphatase

Dye-phosphate
reaction

No reaction

Contents
turn blue

Contents
remain white

*The phosphatase test. The substrate sodium diphenyl phosphate
is added to a pasteurized milk sample, and the tube is incubated.
Then the dye reagent is added. If phosphatase is present, the
tube contents turn blue. However, they remain clear if
phosphatase was destroyed during the pasteurization process.*

■ Figure 38 ■

The protein portion of the milk, the casein, is used to produce
cheese and cheese products. Precipitated from the milk, the protein
curd is an **unripened cheese** such as cottage cheese. The leftover
liquid, the whey, can be used to make cheese foods.

When the cheese is allowed to ripen through the activity of various microorganisms, various cheeses are produced. **Soft cheeses,** such as Camembert, do not spoil rapidly. Camembert cheese is a product of the growth of the fungus *Penicillium camemberti.* **Hard cheeses** have less water and are ripened with bacteria or fungi. Swiss cheese is ripened by various bacteria, including species of *Propionibacterium,* which produces gas holes in the cheese. Bleu cheese is produced by *Penicillium roqueforti,* which produces veins within the cheese as it grows.

Other fermented foods. Other fermented foods are also the product of microbial action. **Sauerkraut,** for example is produced by *Leuconostoc* and *Lactobacillus* species growing within shredded cabbage. Cucumbers are fermented by these same microorganisms to produce **pickles.**

Bread. **Bread** is still another product of microbial action. Flour, water, salt, and yeast are used to make the dough. The **yeast** most often used is *Saccharomyces cerevisiae.* This organism ferments the carbohydrates in the dough and produces carbon dioxide, which causes the dough to rise and creates the soft texture of bread. Unleavened bread is bread that contains no yeast. Sourdough bread can be made by using lactic acid bacteria to contribute a sour flavor to the dough.

The term **industrial microbiology** refers to the use of microorganisms for industrial purposes. Such things as anticoagulants, antidepressants, vasodilators, herbicides, insecticides, plant hormones, enzymes, and vitamins have been isolated from microorganisms or produced in large quantities by genetically engineering the organisms with foreign genes.

Microbial Products

In commercial industrial plants, microorganisms are widely used to produce numerous organic materials that have far-reaching value and application.

Enzymes. Among the **enzymes** industrially produced by bacteria are **amylases,** which break down starches to smaller carbohydrates for commercial use. Amylases are also used in brewing, baking, and textile production. Bacteria have been used to produce **proteases,** which break down proteins and are used for tenderizing meats, preparing leathers, and making detergents and cheese.

Polysaccharides. The food, petroleum, cosmetic, and pharmaceutical industries use microorganisms to manufacture **polysaccharides.** For example, the bacterium *Xanthomonas campestris,* produces a polysaccharide called **xanthan,** which is used to stabilize and thicken foods and as a base for cosmetics. It is also a binding agent in many pharmaceuticals and is used in textile printing and dyeing. Another polysaccharide of microbial origin is **dextran.** The bacterium *Leuconostoc mesenteroides* produces this polysaccharide when it grows on sucrose. Dextran is used to extend blood plasma.

Nutrients. Amino acids, nucleotides, vitamins, and organic acids are produced by the ton by microorganisms. Various types of research and health laboratories use these products, and health-food stores sell them as **nutritional supplements.** For example, the **lysine** prescribed by some doctors to treat herpes simplex infections is a product of the bacterium *Corynebacterium glutamicum.* **Vitamin B$_{12}$** (cyanocobalamine) and **vitamin B$_2$** (riboflavin) are produced by a bacterium and a mold, respectively.

Chemotherapeutic agents. Another valuable use of microorganisms in industry is in the production of **chemotherapeutic agents.** Almost two billion dollars worth of drugs are produced in the United States, mainly by the use of microorganisms. Antibiotics are produced by fungi such as *Penicllium* and *Cephalosporium* and by species of the bacterium *Streptomyces.* Many of these drugs are natural, but several are synthetic or semisynthetic drugs that begin with the naturally occurring molecule, which is then modified.

Insecticides and Leaching Agents

One of the more important uses of microorganisms is as **insecticides.** The bacterium *Bacillus thuringiensis* produces in its cytoplasm a toxic crystal that kills caterpillars such as gypsy moth larvae and tomato hornworms. The toxic crystal damages the digestive tract of the insect and allows local bacteria to invade and destroy the tissues. The bacillus is sprayed on plants during the growing season and is regarded highly for its insecticide quality.

Another bacterium that can be used as an insecticide is *Bacillus popillae.* This organism causes milky spore disease when it infects caterpillars.

In the **mining industry,** microorganisms are widely used to leach low-grade ores to extract their valuable metals. For example, copper and uranium can be leached from low-grade ores by species of *Thiobacillus.* This organism frees the iron ions from ferrous sulfide

by oxidizing ferrous ions to ferric ions. The ferric ions then oxidize copper sulfate to a soluble form which can easily be collected.

Fermentations and Biotechnology

The fermentation of carbohydrates by yeasts to produce ethyl alcohol is used by the alcoholic beverage industry. **Wine** is the aged product of alcoholic fermentation of fruits. The wine-making process begins with crushing and stemming the grapes to produce a product called **must** (Figure 39). Sulfur dioxide is used to kill wild yeast and other organisms, and the must is then combined with species of *Saccharomyces* to ferment under carefully controlled conditions. The skins and seeds of grapes contribute to the color and flavor of the wine. Fermentation takes several days and results in a wine having an alcoholic content of about 15 percent. In **dry wines,** most of the available carbohydrate has been used, while in **sweet wines,** some carbohydrate remains.

An aging process follows in which the wine is placed in barrels to develop its full flavor, aroma, and bouquet. The wine extracts chemicals from the wood. Then it is filtered, pasteurized, and bottled for market, unless secondary fermentations are intended. For example, **sparkling wines** are inoculated with sugar and permitted to continue to ferment. The carbon dioxide produced accounts for the bubbles in a sparkling wine such as champagne.

For the production of **beer,** a type of grain, usually barley grain, is used. It is mashed, soaked, and filtered to produce a liquid called **wort.** The wort is inoculated with *Saccharomyces* and permitted to ferment. The final alcoholic content of the beer is approximately 7 to 8 percent. The beer is pasteurized or filtered before it is placed in cans or bottles for sale.

One of the most elegant expressions of industrial microbiology in the modern era is the science of **biotechnology.** In biotechnology, microorganisms are used to produce medically or industrially important products such as drugs, medicines, and pharmaceutical agents. Bacteria, yeasts, and algae are commonly used for this purpose. Gene

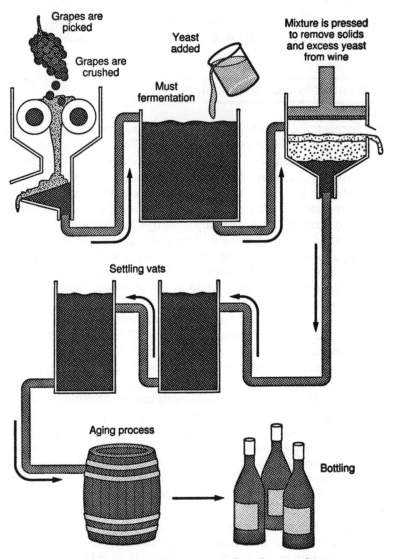

The process used in the commercial production of wine.

■ Figure 39 ■

manipulations are made, and the genetic constitutions of organisms are changed in the process of **genetic engineering.** Organisms are cultivated in huge volumes in tanks called fermenters. The desired substances are produced within these tanks.

Genetic engineering applications began when scientists discovered that they could open DNA molecules at desired points, insert foreign gene fragments, reclose the DNA segments, and introduce them to recipient cells. The cells would respond by encoding the proteins specified by the foreign genes. One of the first products so produced was **insulin,** first licensed for use in the 1980s. **Human growth hormone,** human **interferon,** and other substances such as **clotting factors** were subsequently developed by genetic engineering methods. These products are highlighted in earlier chapters.